THE GOLDEN CONTRADICTION:
A MARXIST THEORY OF GOLD

This work is dedicated to my brave and patient daughter
Aliyah

The Making of Modern Africa
Series Editors: Abebe Zegeye and John Higginson

Manpower Development Planning
Theory and an African case study
Berhanu Abegaz

Essays on Ethiopian Economic Development
Edited by Berhanu Abegaz

Labour Relations in a Developing Country
A case study of Zimbabwe
Mark A. Shadur

Pseudocapitalism and the Overpoliticised State
Reconciling politics and anthropology in Zaire
S.N. Sangmpam

State and Resistance in South Africa, 1939-1965
Yvonne Muthien

An Anatomy of Public Policy Implementation
The case of decentralization policies in Ghana
Joseph R.A. Ayee

Large Commercial Farmers and Land Reform in Africa
The case of Zimbabwe
Peter von Blanckenburg

Tributors, Supporters and Merchant Capital
Mining and underdevelopment in Sierra Leone
Alfred B. Zack-Williams

International Joint Venture Formation in
the Agribusiness Sector
The case of Sub-Saharan African countries
Habte Gebre Selassie

Religious Impact on the Nation State
The Nigerian predicament
Pat Williams and Toyin Falola

The Golden Contradiction: A Marxist Theory of Gold

With particular reference to South Africa

Dr Farouk Stemmet
Independent scholar, Leeds, UK

Avebury
Aldershot • Brookfield USA • Hong Kong • Singapore • Sydney

© F. Stemmet 1996

All rights reserved. No part of this publication may be reproduced, stored in a retrieval system, or transmitted in any form or by any means, electronic, mechanical, photocopying, recording, or otherwise without the prior permission of the publisher.

Published by
Avebury
Ashgate Publishing Limited
Gower House
Croft Road
Aldershot
Hants GU11 3HR
England

Ashgate Publishing Company
Old Post Road
Brookfield
Vermont 05036
USA

British Library Cataloguing in Publication Data
Stemmet, Farouk
 The golden contradiction : a Marxist theory of gold with particular reference to South Africa. - (The making of modern Africa)
 1.Marxian economics - South Africa 2.Gold - Economic aspects
 I.Title
 335.4'12'0968

Library of Congress Cataloging-in-Publication Data
Library of Congress Catalog Card Number: 96-85711

ISBN 1 85972 353 5

Printed in Great Britain by the Ipswich Book Company, Suffolk

Contents

Figures and Tables   vi
Acknowledgements   vii
Preface   viii

Introduction   1

Part One: Towards a theory of gold   23

1 Gold as money   27
2 The money-commodity and the value of labour-power   49
3 Gold as capital   77

Part Two: Gold production, capital and the value revolution of the 19th century   99

4 The changeover from placers to lodes   103
5 Universal industrialisation and social capital   115

Part Three: The value of labour-power in early South African gold production   145

6 Competition in early Witwatersrand goldmining   149
7 The black goldminers and the 'fixed gold price'   187

Conclusion   247

Bibliography   255
Index   269

Figures and Tables

Fig. 1: Section through typical deep mine   9
Fig. 2: Share-ownership structure of the group system in
         Witwatersrand goldmining   168
Fig. 3: Competition and the group system in early
         Witwatersrand goldmining   172

Tbl. 1: Total world precious metals production, 1741-1910   105
Tbl. 2: Gold finenesses and their uses   124
Tbl. 3: Witwatersrand and the black mining population,
         1889-1904   232

Acknowledgements

My sincerest thanks to Hillel Ticktin, who supervised this PhD, for stimulating debates over the three years. Hillel dealt with my philosophical sloppiness by patiently allowing me to maintain an idea until I discovered its weakness for myself. He has always been respectful of my views. I hope that I have reciprocated.

Paul Trewhela, the first person with whom my thoughts found resonance, played a central role in the formulation of my ideas. He has also generously lent his unpublished manuscripts to me. This study has also benefited from discussion and debate with Baruch Hirson, Bill McElroy, Alan Horn, Paul Smith, Peter Kennedy, Ian Spencer, Janet Campbell, Alun Francis, Katrin Querfeld (who also helped with my German), Julian Siann and Russell Ally in Britain and with Kevin French and Richard Harvey in South Africa.

My meticulous proof reader, Richard Maudsley, helped at very short notice. Any errors that might remain are my own. Kevin French and Lindi Woolly generously made their home available to me in Johannesburg. Also Bill McElroy in London and Colin Buchanan in Glasgow opened their homes and hearts to me. I thank the kind and helpful staff of the Rothschild Archives, London, the Mineralia Library and the Chamber of Mines Library, Johannesburg and the Institute of Social History, Amsterdam.

This study was funded by an Africa Educational Trust Scholarship — flexibly administered to allow for the early purchase of a laptop computer, thus enabling completion within three years. My sincerest thanks to the Trust and, in particular, to Jill Landimore.

The staff and headmaster of Lomond School, Scotland, and in particular the Burnbrae mob, have provided a warm and supporting home in which to live and work. Regular afternoon tea with the matron, Cary Alexander, was my lifeline with reality. My daughter, Aliyah, has been most tolerant of my long periods of absence from her. I have asked a lot of this child and have received without question. I shall remain forever grateful to her.

Preface

A work of political economy looking at the heyday of 'hard money', *The Golden Contradiction* asks how the traditional 'constancy' of gold changed into a daily-fluctuating gold price. Its central thesis is that gold materialises the contradiction between money and capital. The fixed gold price is radically re-assessed. Gold-producing labour is shown to have operated according to different economic rules to those for labour in general. This underlies both the absurd life of the 'Wild West' gold digger and the origins of *apartheid* in South Africa.

California from 1849 to c.1865 and South Africa from 1886 to c.1907 are set against the backdrop of a world history of gold production to reveal a revolution in gold economics. Far from South Africa's first goldmining companies initiating a separation of black from white mine workers, it is argued that the mines took advantage of a division already in place and sought, instead, to separate black goldminers from black workers in general. This was the most elementary form of *apartheid*.

Using Marx to go beyond Marx, it celebrates the elegance of theory as a process. Part theoretical, part descriptive and part polemical, it had a long gestation period, having been dismissed by orthodox economics professors as a non-starter. Eventually submitted as a PhD thesis to the University of Glasgow in 1993 under the title *Control of the Value of Black Goldminers' Labour-Power in South Africa in the Early Industrial Period*, it appears here largely unaltered. Comprising the first part of a general theory and modern economic history of gold, it takes the question up to the point where gold production by joint-stock companies becomes the dominant form and metallic gold ceases to circulate as currency. Still awaiting the author is a study of the role of this metal in the age of 'derivatives'.

Introduction

When there is a general change of conditions, it is as if the entire creation had changed and the whole world been altered.

—*Abd al-Rahman Ibn Khaldun*

The object

The revolution which had taken place in the value of gold, the measure of value, in the second half of the nineteenth century, coincided with the need of international trade to hold fast the value-ratio at which the world's various paper currencies represented a definite weight of gold. Movement in value did not translate into movement in price. Therefore, in order for this metal to be produced at all, even on the basis of simple reproduction, its value would have had to be lower than that value represented by its secular price.

The 'fixed price' of the product also acted as a value-barrier between goldmining and industry in general. This meant that goldmining had to be responsible for its own surplus-value production, rather than, as does other industry, claim it from the general pool of surplus-value on the basis of the volume of capital invested. The size of its own *variable* capital, therefore, came to have an immediate bearing on its rate of surplus-value. For gold-producing capital, the value of gold-producing labour-power becomes crucial to its function as capital. For gold production to be secured, appreciation in the value of gold-producing labour-power needed to be controlled. The early South African goldmining industry offers the clearest demonstration of this. The object before us, therefore, is the value of labour-power in gold production.

This treatise offers three things: (i) a contribution towards a theory of gold (ii) a reassessment of the great transformations that took place in the production and political economy of gold in the second half of the nineteenth century (a 're-appropriation of the material in detail'); (iii) an alternative explanation for the emergence of oppressive labour conditions on the South African gold mines. Thus is the presentation divided into three corresponding sections.

Statement of method

Writings on the political economy of gold have been virtually monopolised by liberal economists since the time of the international gold standard system in the 1870s. This historiography has been erratic, consisting of floods of material at times of disquiet in the world economy, punctuated by periods of virtual silence. For example, writings in the 1880s and 1890s were directed at the gold standard and value-ratio between gold and silver. The next wave, from the end of the First World War to the 1930s, was concerned with the reinstatement of the collapsed gold standard and whether national currencies should be pegged to gold at the pre-war level. A trickle of literature links this period to the near-panic turnout of the 1960s and early 1970s. The relationship between the United States dollar and gold was here the principal issue. This was followed by another burst of material in the early 1980s when economists were dispairing over apparently uncontrollable inflation. From time to time there appeared re-evaluations of earlier periods or themes. This voluminous body of literature is as much the work of academics as it is of finance ministry officials and practical banking men. There have also been a number of journalistic overviews.

All this literature starts from the received assumptions of neo-classical economics. Alternative writing on gold, especially Marxist writing, is conspicuous by its near-absence. This latter historiography is introduced by Friedrich Engels' remarks in Volume II of Marx's *Capital*. There are some descriptions of South African gold production in John Hobson's *Imperialism*, while V.I. Lenin's remarks on gold are known more for their graphic imagery than for their analytic content. Ernst Mandel's *Decline of the Dollar*, which appeared in 1972, is a collection of essays he had written between 1964 and 1972. In the same year Suzanne de Brunhoff published *Marx on Money*, a work which makes no reference to Marx's *Grundrisse*. In 1981 the journal *Capital and Class* carried the article *Capitalism and Gold* by Duncan Innes. Anikin's *The Yellow Devil* was translated from the Russian in

1983, the same year as Gool published his critique of South African goldmining historiography, *Mining Capitalism and Black Labour in the Early Industrial Period in South Africa.*

Understandably, South African writers have made a disproportionately high contribution to world scholarship in this area. This historiography has undergone its own evolution (Gool, S., 1983). An early radical interpretation of South African gold production is offered by Leo Katzen in his *Gold and the South African Economy,* published in 1964. Much material exists as articles in journals (some of which are obscure), while many others are unpublished theses, such as Paul Trewhela's thoughts on the gold question, committed to paper in 1981 and 1986. In 1975 Michael Williams offered his theory of gold in an obscure London publication, *Journal of the Conference of Socialist Economists.* A number of articles by Peter Richardson and Jean-Jacque van Helten have appeared since 1977. Russell Ally's thesis, *The Bank of England and South Africa's Gold,* was lodged in 1990.

South African left wing historiography on gold suffers from two particular weaknesses. Firstly, apart from Williams, Trewhela and Innes, there have been few attempts to construct a political economy of gold. Theory, such as there is, consists either of quoting Marx (e.g., Williams), proposing arbitrary categories (e.g., Wolpe), or the academic project is dictated by an *a priori* objective to 'place the African working class more in the centre of the stage and attribute to its struggle a more central role' (Gool, op. cit., p. 16).[1] If theoretical exploration cannot satisfactorily answer the question, 'how does this inform the struggle?' then the exercise is scornfully dismissed as irrelevant.

Secondly, although there has been polemic, the writing does not constitute a debate.[2] Every writer appears to have claimed a niche for himself into which he delves ever more deeply for original material. This, of course, has a very important role. But it is our view that while such work adds to information, it contributes less generously to understanding. Indeed, this division of labour and non-interference in one another's 'patch' appears to rest on an unspoken *quid pro quo*. It has become an element of leftwing academic etiquette to acknowledge a writer's expertise if he has scoured an archive for original papers. When reference is made to work in another detailed area, conclusions are taken as proven and interpretations are taken to be the most representative of reality. To rework material already studied by others is seen as a waste of labour. Thus is the theoretical poverty both concealed and protected.

These two weaknesses result in a historiography which is not only empiricist (albeit left-empiricist), but also parochial. The explanation for the history of wages on the South African gold mines, e.g., comes down to the high cost of producing gold from low-grade ore (in the context of a fixed gold price) which the mines have managed to do because they have successfully split the working class along racial lines and incorporated the white workers into the control structure of the mines. This very formulation states that the 'splitting' of the working class is a particular local solution to a world problem. But this world problem is never examined, although briefly asserted by Ticktin. The only possible relationship which can be allowed to exist between any particular country (especially a colony or ex-colony) and the world in general (but especially those countries possessed of large-scale capital) is 'imperialism'.[3] 'Monopoly capitalism' is a similar article of faith. Once this has been stated, theoretical obligations have been fulfilled and the scholar can get on with the task of examining new material in order that 'the struggle' may be 'better informed'. In other words, where a global context is acknowledged, this is either done uncritically or it is only invoked in the service of telling the story of *South African* capitalism. How can one explain South Africa without being able to explain the material singularly responsible for its existence as a modern state? At the same time it would be a mistake to see the need for a theory of gold only in terms of explaining South Africa. The position should rather be the reverse, with South Africa illustrating the workings of such a theory.

The category around which this dissertation is constructed is the value of labour-power. This means that the value of labour-power is used as a probing device in the service of thought. All manner of disparate facts settle into a coherent relationship if we keep our eyes focused on the value of labour-power. But why is this not as arbitrary as, say, 'focusing on' monopoly capitalism? What consistently emerges from analysis is a question of whether or not and, if so, then how, gold-producing labour reproduces itself. The material examined in detail is not only South African, and not only from the turn of the century. The factual base of this thesis covers antiquity, the classical world, the Middle Ages, pre-colonial Africa and South America, California, Virginia, Colorado, British Columbia, Alaska and the Klondike, Australia, New Zealand, India, Siberia, joint-stock companies, the industrial revolutions, the abolition of slavery, the technical aspects of mining in general and goldmining in particular, the evolution of money (both historically and as a category) in addition to the history of southern Africa from c.1870 to c.1910. This list is declared only by way of saying that the use of the category value of labour-power is not a case of '*a priori* positionalism', to borrow Eddie Webster's lovely phrase. The material examined suggested its own category.

The value of labour-power is not a 'Marxist' category. It is used in many different places, although not always by the same name (see, e.g., the neo-classical economics essay Bonacich, E., 1972). Our project deals with the relationship between the value of labour-power and the production of the material gold. Our problem thus transcends both that of the money-commodity and that of capital. We have conducted our investigation according to the method employed by Marx in *A Contribution to the Critique of Political Economy* and the *Grundrisse*, and explained in the latter. Our understanding of the nature of labour-power, commodity, money, capital, etc. derives from these two works plus Marx's *Capital*. Having said that, we must state that we have deliberately shied away from simply quoting from Marx as 'proof' of our assertions. Instead of this we have attempted to *derive* our propositions for ourselves. We have, after all, the benefit of a large body of material not available to Marx. Most of the time our results coincide with those of Marx, but sometimes they do not. Where they do not, we have tried to show why we think Marx in error. We think it necessary to state this because there are scholars for whom critique of Marx borders on impertinence. Since we do not subscribe to critique by quotation, it is up to those who understand Marx better than we do to demonstrate our shortcomings and thereby improve our understanding.

In this work, a clear distinction is maintained between gold as use-value and gold as money and between gold as money and gold as capital. At every stage of the investigation these are first considered in isolation from one another, and then in their interrelation. This is methodologically different from what appears to be the practice in the literature where forms are reduced to static elements independent of one another. Labour, product, commodity, money and capital are not understood as forms. Form A which posits form B is itself, in turn, posited by form B as form A1. Although labour posits commodity production, under commodity production labour itself comes to take a different form to labour which has not posited commodity production. Similarly the commodity before and after it has posited money, or money before and after it has posited capital. Labour may posit capital, but capital posits labour as wage-labour. So, too, is the commodity which arises out of capitalist production a different form of commodity to that which gives rise to capitalist production. This distinction between whether a form posits or is posited is very important to our thesis, especially since it is the self same material, gold, which is at the same time product, commodity, money and capital, without, in essence, ever having to change its *physical* form. To begin to understand the complex role of gold in the modern world, it is necessary to be clear on when this substance presents as commodity, when money and when capital.

Methodologically, therefore, one is required first to consider gold simply as a thing, to appreciate it as a concentrate of physical attributes and to know the material/technical character of its production. Then it must be viewed as a commodity, subject to all the laws applicable to commodities. The subsequent question then becomes one of understanding how these laws express themselves in a commodity which has isolated itself from the general body of commodities and stands opposed to it as money. In other words, how do the laws of commodities translate into the laws of money, and similarly, the laws of money into the laws of capital? While the social form of this objectified labour takes all these different forms, the material form of that objectified labour, a lump of gold, remains the same. This means that the same material presents simultaneously in all these different social forms. But the commodity is, by definition, a contradiction, and money and capital are developed moments of that contradiction. Gold, in its social character, must therefore demonstrate all of these. The labour which produces commodities is posited, socially, as commodity-producing labour. Similarly, the labour which produces money is posited as money-producing labour, and that producing capital, capital-producing labour (wage-labour). Gold is, simultaneously a concentrate of this contradiction at all its moments. A concern of this thesis is to examine the character of that labour which objectifies itself in this socially polymorphous way.

The period covered in detail by this thesis begins with the opening of the Californian goldfields in 1849 and ends with the repatriation of Chinese indentured labourers from the Witwatersrand in 1907, although occasionally reference is made to situations falling outside of this period. During the latter half of the nineteenth century gold production changed from being based predominantly on placers to being based predominantly on lodes. The one-man digger was replaced by the joint-stock company. Capital requirements changed from pick, shovel and pan to hundreds of thousands of pounds. Slavery was replaced by wage-labour. The industrial revolution (eventually) came to mining and, in the overall mining production process, the balance between underground work and surface work altered very strongly in favour of the latter. In short, there was a revolution in gold production.

The earliest comprehensive account of these developments, as far as we have been able to ascertain, was Alfred Lock's 1,229-page *Gold: Its Occurrence and Extraction*, which appeared in 1882, one year before Marx's death.[4] There have been major treatises on the gold question in the wake of the Californian and Australian discoveries, but these tended to speculate on the probability of the *price of gold* retaining its traditional stability. What concerned scholars at this time was the dramatic rise in gold *output* from the increasingly scientific working of the placers, rather than the eventual depletion of those placers and the re-organisation of gold production on a different value-basis. All the authorities whom Marx could consult, fell within this category.[5] Marx's treatment of the gold question is inadequate for the purpose of constructing a political economy of this material. But this is not only on account of his sources, but also because the political economy of gold was not his project. His object was to lay bare the laws governing production on the basis of capital.[6] To this end, a detailed political economy of the production of gold would serve no purpose. His most important insight into gold is the nature of surplus-value extraction where gold has already become money. This is part of his exposition of the evolution of exchange value: to wit, its assumption of an independence from particular commodities. But on the basis of Marx's understanding of exchange value alone, one would not be able to explain how it is that 'when the object [of production] is to obtain exchange value in its specific independent money-form', but this is done by free labour or free wage-labour, the 'recognised form of over-work' is *not* 'working to death' — contrary to his conclusion in *Capital*. We will show that many of Marx's utterances on gold are not particularly helpful in constructing a political economy of this metal.

This book also sets out to prove that the infamous labour practices on the South African gold mines are a direct expression of the developments introduced above, both in terms of the social evolution of money, and in terms of the technical revolution in gold production. We suggest that the existence of a low-value, low-productivity labour force *in the region* offered goldmining capital an opportunity to circumvent the absence of low-value, high-productivity production in goldmining in general. Other goldmining areas around the world at the time were not so 'fortunate'.

Some general points on gold occurrence and extraction

The history of gold production and technical questions about its geological occurrence and its mining and processing do not generally fall within the remit of the social scientist. Many of the discussions which will follow below require some familiarity with such historical and technical questions. For the benefit of the reader not familiar with these aspects of gold production, we offer these general points by way of introduction.

Gold occurs on or near to the surface of the earth in two principal conditions: (i) the first of these is *placer deposits* (commonly and inaccurately refered to as 'alluvial'[7] gold). These are deposits which originated elsewhere and at a later stage ended up 'placed' in their locations, mainly by movement of water, but also by movement of wind and sand. Since they are relatively younger than their matrix, they are not geologically integrated with it and hence relatively easy to extract. Such gold is usually native gold, i.e., in metallic condition, thus restricting necessary processing to physical extraction.

Not all placer deposits occur on the surface. Deposits created during past geological era lie buried deep beneath many layers of more recently formed strata and hence are called *deep placers*. This is particularly so beneath modern gold-bearing river beds. While the working of placer deposits, therefore, mainly involve surface work, an element of both open cast and underground mining is also associated with this kind of deposit. Extracting such underground deposits involves the additional difficulties and complications of underground mining, but the gold remains geologically distinct from its matrix. Placer deposits are characterised by high yields of gold per unit ore and relatively quick reward for effort. Almost all gold mined up to the third quarter of the nineteenth century occurred in nature as placers.

(ii) The second condition of naturally occurring gold is as *lodes*.[8] Lodes differ from placers principally in that lodes are sites of the original formation/concentration of the element at a very early time in the formation of the earth. Such gold occurs in a condition where it is integral with its matrix and thus not amenable to segregation (or, often, detection) by simple physical means. The lodes themselves occur either in fissures which occasionally puncture the surface as outcrops, or as massives, being simply 'bubbles' of gold-bearing material suspended in other rock. The matrix itself is often quartz and hence extremely hard compared to placer matrixes. The gold particles themselves tend to be minute, often not visible to the naked eye and their segregation from the matrix involves chemical treatment of the ore after the latter had been sufficiently pulverised to allow for adequate contact with all its particles.

(iii) Conglomerates, a hybrid of placers and lodes, is the third condition in which gold occurs naturally on earth. These are placers of such ancient age that they pre-date the final violent geological upheavals in the formation of the earth's crust.[9]

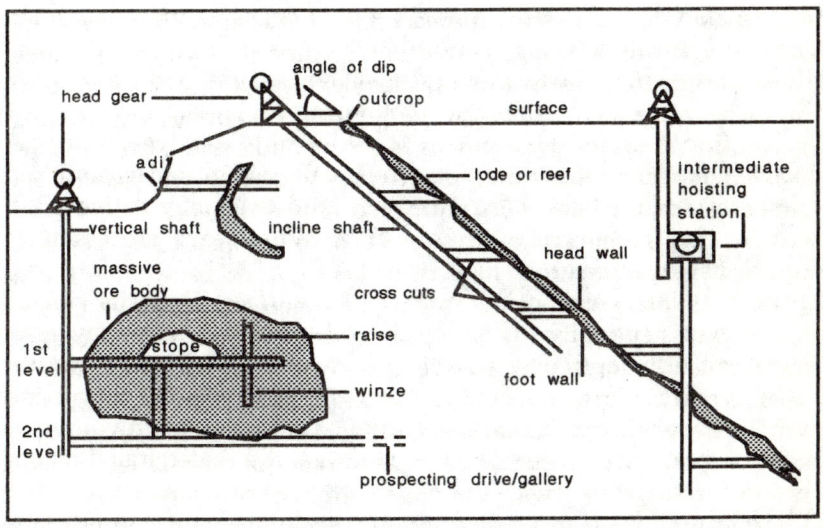

Figure 1   Section through typical deep mine
Source: Simplified from Encyclopædia Britannica, 1982, Vol. XII, p. 248

The original placers were put through a process not unlike that of lode formation. Under conditions of the most extreme heat and violent pressure, the gold and its matrix got fused into an integrated quartz body, so that, although strictly speaking a placer, the condition of the gold within its matrix resembles that of a lode. Such gold is described as occurring in quartz lodes (also described as reefs). The gold deposits of the Witwatersrand fall into this category. For the purposes of this study, therefore, conglomerates may be regarded in the same way as lode gold.[10]

The first feature of placer mining to strike the scholar investigating its history, is the remarkable constancy of its labour process. This process has been virtually unaltered for millennia and occurs in much the same form the world over. A vessel (pan, calabash, batea, etc.) is used for scooping up ore which is then 'washed' in water by the water and lighter particles being allowed to spill across the edge through a steady rotary motion of the arms, relying on centrifugal force to restrain the gold particles, which are more massive, within the vessel. In many cases this work is divided amongst a team, the latter often consisting of men, women and children. The furthest development in placer mining by the mid-nineteenth century appears to be the channelling of water-borne ore through troughs or ditches equipped with transverse riffles to trap the slower moving heavier particles. Such a system was, however, also in use in ancient Egypt (Lock, op. cit., pp. 6-10).

The first, 'surface work' involves sifting and prising from placer deposits either on or within about five metres of, the surface. At its most sophisticated level in antiquity, this involved a series of holes in the ground, in other words, a lowering of the surface, rather than a puncturing of it in order to work beneath it. From about 1900, large-scale placer mining has been carried on by means of dredgers, especially in the USSR. Except where the ore occurs from surface to depth over a wide area, the hole sometimes simply get bigger and bigger, eventually turning into a pit or an open-cast mine. Surface work is the earliest and most simple mode since it can be carried out with bare hands or absolutely minimal mechanical aids and is associated with the most primitive of peoples.[11]

The second mode, 'deep digging', or 'mining' (as understood until the 1890s and not to be confused with the 'deep-level mining' of Witwatersrand fame) also involves placer deposits. This time, however, they are located too deep underground or within mountain crevices, thereby making surface work impractical. Almost all the gold mines from antiquity to the mid-nineteenth century were of this kind. Since the deposits were placers, the matrixes in which they occurred were invariably relatively young and consisting of material which was relatively recently disturbed. There is no 'organic' bond between the gold and its surrounding material. It is therefore amenable to extraction with the aid of hammers and chisels or pick-axes. This mode is thus dependent on the art and science of metal-working and metallurgy already having been achieved. Since the technology of propping and — equally importantly in some cases, as in ancient Egypt — the material for propping were not available, the shaft with galleries or stopes developed much later than mining in mountains, where the rock supported itself, although the former, too, have been in use for centuries.[12]

Once retrieved as ore, the gold particles need to be separated from the non-gold particles. This usually involves some manner of washing in water. In the case of placer gold this process is in most cases identical with that of extraction, since the gold already occurs as either nuggets, grains, flakes, specs or dust, or, if embedded in rock, the latter occur as pebbles of various sizes requiring first to be pulverised by means of hammers. Because gold is one of the heavier metals, it can, by the motion of water, be segregated from other, lighter material. At its simplest level, this involves panning, as described above. Whether amongst Ashanti women on the Gold Coast, Californian Forty-Niners, or Australian Fossickers, panning involves the identical process.

The most important points about placer mining relevant to the construction of a political economy of gold, though, are, (i) that gold occurs in nature in a ready-to-use metallic (native) form; (ii) the historical and geographical constancy of production technique; (iii) the, on average, fairly constant yield per unit labour expended; (iv) the fact that placer mining can be carried on by single individuals working as isolated units without any form of co-operation between them; and, (v) that this labour may be carried on using only the most primitive of technical aids. These five points are most important in securing for this material its money role eventually in all commodity economies, as we shall see below.

Though not abundant, the natural occurrence of gold on the earth's surface is fairly widespread, so that the initial form of 'extraction' involved little more than sifting and prising for the gold to be claimed from nature.

The technical constraints on underground and lode mining

It is, however, in the technical nature of mining, all mining, but especially of ore-based metals, that the principal concrete labours particular to it involve hammering, digging and hauling (and washing). This has been a feature constant to it from the earliest mining up to the present century. The first purpose here is to discuss these *technical* constraints and show how mining, by its nature, is different from other industries. The limited object being to ensure that what is peculiar to mining in general is not assumed to be peculiar to goldmining in particular. We are interested in these features in so far as they distinguish mining labour from other labour, on the one hand, and goldmining labour from other mining labour, on the other. These may be listed as:

(i) It is almost always the case that the site of mining production is in an inaccessible place — such as deep underground. Extractive industries offer the producer no choice of site of production. The material has to be extracted from where it had been deposited by the chance forces of nature, even though a greater or lesser degree of final processing can take place at a site removed from the primary and more to the producer's convenience.[13]

(ii) Those sites closest to settlements will, naturally, have been the first to be exploited, i.e., earlier in the evolution of technique, rather than later.[14] These developments, in some instances, impeded the conveyance to the site of whatever technology was available and the servicing of that technology and the workforce on site, be it by man, beast or the elements. E.g.:

> Difficulties encountered [in opening up the Tati goldfield, Matabeleland], were the want of pumping machinery, and the great cost of transport from the coast (ibid., p. 15);

> Ramses ...turned his particular attention to the gold-districts which had been discovered... But water was wanting in the dreary sterile valleys of the mountainous country, and men and beasts died on the roads to the gold-districts (ibid., p. 6).

In other instances, these developments technological development brought more distant and formerly inaccessible resources within reach of exploitation. The relation between technological development and mine accessiblity is therefore uneven and case specific.

(iii) Often the site itself did not admit anything larger than a crawling man, if that. Labour spent removing superfluous material is labour lost from the rockface. Tunnel cross sections are therefore kept to an absolute minimum, even where not technically thus constrained. A premodern mine had no room in which to manoeuvre anything much larger than a man, let alone whatever he might require to operate it. Coal mining in Victorian Britain offended even the gentlefolk with its images of wretched children crawling along galleries harnessed to loaded cocopans.[15] Mining is therefore compelled to forsake a great many technical developments generally available to other industries, restricting itself to the bare minimum means without which production could not take place — the labourers equipped with hand-tools.

And by way of proving that this is not a case of social form, one might select examples from a few places very different to the country at the forefront of nineteenth century capitalist development, e.g., Stalin's Russia:

> The tunnels inside the mines were so narrow that two people could hardly walk side by side, and the height hardly permitted a medium-sized fellow to walk unless bent from the middle. In some places the gallery was so badly excavated that we had to advance on our knees and hands (Conquest, R., 1978, p. 122);

and Tacquah, in pre-colonial West Africa:

> The head- and foot- walls are composed of syenite, as hard as flint, through which the natives are unable to penetrate, so that, unless the reef itself is wide enough to allow a man to work in it with elbow room on each side, they can do nothing with it (Lock, op. cit., p. 32).

(iv) The basic actions of hammering, digging and hauling, concrete labour which characterised mining through the ages, are actions of brute force. The application of concentrated force at a specific point, until relatively recently, required the congregation of a large number of discreet force units within the proximity of that point. The two hundred slaves required to haul one granite block themselves took up an area much larger than the block which they hauled. By the eve of the introduction of steam ships, sailing ships, by then at their largest and carrying greater than ever cargoes so as to make oceanic voyages as economically viable as possible, could hardly be seen for their sails. Coaches and wagons could convey heavier loads only by the addition of more beasts.[16] Prior to the invention of bellows, the concentration of great heat in a small area so as to enable smelting was a serious problem. The solutions, though sometimes ingenious, always involved an area many times larger than that of the material worked on.[17] The point here being that concentration of force could only be achieved where space permitted the congregation of a large aggregate of relatively small discreet force units. Mining, therefore, by its very nature, had to wait for the invention of electricity (or, at least, steam power), pneumatic tools and dynamite, before force greater than can be mustered by a man equipped with hand tools could be systematically introduced.

The crudeness of mining technology can be illustrated by many examples from around the globe and from all ages. 'Until late in the 1950s', says Conquest, 'prisoners were still working with picks and axes as their only tools' (Conquest, op. cit., p. 122). These words, describing the Siberian prison mines at Kolyma, could have been said about almost any mine, whether prison or not, Stalinist or not. And in the gold mines of the pharaohs: 'the strongest men worked in the tunnels, breaking the rock with iron hammers, while the old men and children carried the ore to the crushers, where it was broken up by the men and then further ground by women' (Observer, 02.07.89). And, staying with ancient Egypt, 'Thereupon [after opening of the new gold mines by Seti] everything was done to carry on gold-washing to success. The people who followed this laborious occupation were placed under the supervision of a *hir-pit* or "overseer of the foreign peoples"' (Lock, op. cit., p. 6).

Mining is, therefore, an industry which could only benefit from technological advancement once technique in general had developed beyond a certain level, that level being for mining much further along the line of general technological development than for other industries.[18] Mining could, for all this time, only achieve increased production through the intensification of labour. More material could be broken from the rockface only by slinging more human bodies at it. Mine labour is or was, therefore, already on account of this circumstance alone, brutal and oppressive.[19]

(v) But it becomes even more so, when one looks at points (ii) and (iv) in their mutual interrelation. Given that over time gold became increasingly inaccessible, both through being mined at greater depth, and through new fields being developed at greater distance, labour requirements increased simply to retrieve the same amounts of gold. While this has the effect of increasing the value of gold, it also serves to push production to the technical threshold beyond which no increase in labour input could grant access to known reserves. This can be either because the gold is simply located too deep underground, or in too inhospitable a terrain, or is locked into a particular variety of ore for the nature or yield of which retrieval processes had yet to be developed.

Placer mining (gold-washing) and lode goldmining compared

If placer mining had remained virtually unaltered for centuries, the same can be said for lode mining. Ore bodies, which may also be of the massive variety (see Fig. 1), were reduced by being set about by slaves or other forced labour equipped with hammers, chisels, picks and shovels. Once retrieved, the ore had to be crushed and pulverised. This, again, was a matter of sledge-hammers and brute force. The pulverised material was then treated with mercury in a process known as amalgamation. Further processing retrieved the gold, but much was lost.

In lode mining access to the site of production is usually admitted only at one point: the shaft head.[20] Placers admit diffused access in that gold may be retrieved, e.g., all the way along a river, both from its bed and from both banks including a certain distance up the valley walls. Gold may also be sought in the estuary and a certain distance into the sea and along the beaches adjoining the estuary. Upstream gold may be found all the way to the lode from which the placer originates.

The phenomenon of a *gold rush*,[21] therefore, takes on a completely different meaning depending on whether the gold in question occurs as lode or placer. Placer gold rushes are characterised by a possessed mass of countless Godless individuals all feverishly outdoing one another to stake claims over a large area. Most of the time this involves no cost other than that of a fearsome weapon.[22] The loss involved in the loss of a claim — whether voluntarily through abandonment or sale, or involuntarily (as was more likely to be the case) through economic or physical coercion — is little more than the labour-time expended on it.

When lode gold triggers a gold rush, the only feverish activity likely to occur is on the stock market as various associations for gain seek to establish themselves as joint stock companies. Indeed, gold rushes associated with this kind of deposit is a phenomenon presupposing the existence of a capital market. Gold occurring in this form is almost always inaccessible to the individual and prior to the existence of joint-stock companies only exploitable by the state, whether directly on its own account (Russia, Rome, etc.) or indirectly through, e.g., chartered companies, sponsored expeditions (Latin America), etc. Only bodies such as these disposed over the necessary coercive force to secure labour and keep it working under necessarily brutal conditions. The degree of technological development did not allow for the worker prising quartz to be a free agent.[23]

In the case of placer gold rushes, possession of the land invariably takes precedence over ownership. One of the symptoms of gold fever is that the sufferer is not overly concerned with the niceties of title deeds and the like, as the illustrious John Augustus Sutter learned at the cost of first his sanity, then his life (Cendrars, B., 1982).[24] Because of the enormous amounts of capital involved and the time the owner will have to spend on site before any gold is recovered, ownership of a lode source becomes as important as possession. The mine owner is thus often involved in a mass of fees, licences, permits, taxation, insurance, leases, etc., not to mention purchasing of adjoining lands in case the ore body extends beneath the boundary.[25] All this apart from the securing of long term supplies in what is often an inaccessible area, and the additional expenses of underground exploration. The extent, grade, location and nature of the ore body needs to be fairly exactly established not only to secure capital, but to keep much capital from drifting off to more profitable ventures.

While in the case of placers the gold occurs in native form, that in lodes constitutes an ingredient of a fused gold-bearing ore. There is, therefore, much less gold in a unit of lode ore than in the same unit of placer ore. In placer mining the gold can be available from the moment someone cares to pick it up. In lode mining there is a necessary time lag between labour first being applied, and usable gold emerging. Lode yields are, therefore, always much less than placer yields, whether considered by unit tonnage of ore, time input or labour input. Historically, much less mining took place by this method than by the placer method, so that the socially necessary labour-time for the production of gold was set by placers.[26]

Since the end product of both methods is the same homogeneous material, lode mining was therefore necessarily much more brutal than placer mining. The distinction between placer gold and lode gold thus lies not only in the technical difference, but in the different ways in which they are significant in a political economy of gold.

Notes

1. Thus is any action on the part of labour which does not coincide with the interests of capital automatically put down to 'resistance'.
2. An exception to this is the debate triggered by Blainey's *The Lost Causes of the Jameson Raid* in 1965.
3. Thus, e.g., are an investigation into the transfer of capital between 'the imperialist countries' and South Africa; or the machinations of the British Colonial Office in the region legitimate areas of study.
4. Marx also appears to have been unaware of the US Government's *Gold Panic Investigation Report* of 1870.
5. Marx's sources include Jacob, W., 1968a and 1968b, both originally published in 1831; Chevalier, M., 1859; Soetbeer, A., 1879, etc. Soetbeer's reticence is surprising, given that by the 1870s, placer production was already revolutionised. Most of the information available to Lock in 1881 would also have been available to Soetbeer in 1878.
6. 'In this work I have to examine the capitalist mode of production, and the conditions of production and exchange corresponding to that mode' (Marx, 1983, p. 19, Preface to the first German edition).
7. Alluvial gold is a particular variety of placer gold. Placers may occur in a number of varieties, of which stream (alluvial) placers have historically been by far the most important (e.g., Klondike). Eluvial placers are distinguished from alluvial placers in that they are not created by streams, but by rainfall and wind, which carry away the lighter materials, thereby concentrating the gold (e.g., early Australian deposits). Beach placers are formed by the different rates at which heavier and lighter materials are shifted by wave action and shore currents, thereby concentrating them (e.g., Nome gold, Alaska). In the case of eolian placers, the principle of concentration is the same as in all other varieties of placers, except that the wind acts as concentrating agent removing fine particles of the lighter dross (e.g., some Australian deposits).
8. Not to be confused with *veins*, of which lodes are but the metalliferous variety. Naturally, geologists would talk of veins, while miners would talk of lodes.

9   Opinion is divided as to the setting-down of the conglomerate beds.
10 Gold is also recovered as a by-product of base-metal mining.
11 Drawing on the letters from West Africa of the German explorer Bosman, Lock says that, 'gold is found in three sites. The first and best was "in or between particular hills"; the negroes sank pits there and separated the soil adhering to it. The second 'is in, at, and about some rivers'. The third is on the seashore, near the mouths of rivulets, and the favourite time for washing is after the violent rains' (ibid., pp. 26-7). In British Columbia, 'the gold is in large particles, and is obtained by the Indians in crevices among and beneath the stones in the river' (ibid., p. 51). Further, in the eastern Transvaal, one prospector reports that, 'a body of Kafirs were at work breaking out quartz, in which the free gold was easily visible' (ibid., p. 20). 'The spots containing the metal are known by the bare and barren surface. The natives dig in any small crevice made by the rains of the preceding winter, and there find gold-dust. These potholes are rarely deeper than 2 or 3 ft.; at 5 or 6 ft. they strike the bedrock. In the still portions of the rivers, when they are low, the natives dig for nuggets that have been washed down from the hills. Sometimes, joining together in hundreds, they deflect the stream, and find extensive deposits' (ibid., p. 15), and in Mexico, 'Glennie found a number of Indians suspended by ropes from the [mountain] crests, picking out the earth containing the gold with wooded stakes' (ibid., p. 110).
12 According to Scherzer, 'the date when the mines of Yuscaran [in Honduras] …were first opened, …is not on record: many of them appear to have been worked for centuries. Five are now (1857) systematically worked. The Malacate mine, the oldest in Yuscaran… was, about seven years ago, taken by a company, who expended great sums in clearing away the rubbish by which the galleries were choked, in order to get to the principal lode. But after 7 seven years of continued digging… the point has not yet been reached where the former miners left off working', (Lock, A., 1882, p. 100). 'La Mina del Oro [New Mexico]… is certainly very ancient… the work was done before …1680. The principal shaft is over 200 ft. deep' (ibid., p. 179).

13   Liquid material, such as oil, offers less site restriction. This is especially the case with off-shore production, where a single platform can serve a number of widely dispersed well-heads delivering to it by undersea pipeline. It must be remembered, though, that liquid extraction, except for water from wells, is a relatively modern industry.

14   Seventeenth century Cossacks hoped to encourage the Russian government, 'to establish those needful posts and centres of supply and trade, the demands for which increased with the wealth which these brigands had obtained through plunder and placer mining', (Del Mar, A. 1969, p. 377).

15   Prospective South African mine workers are put through stringent psychological tests to ascertain their susceptibility to claustrophobia.

16   Hence 'horse power' as a unit of motive force.

17   'The [North American] Indian method of smelting these metals was one of the most remarkable devices of savage ingenuity; ...Having first hollowed out a flat stone to form a basin, they filled it with charcoal, and upon this laid the nuggets of metal. A number of Indians then seated themselves in a circle around the basin, each one having in his hand a long reed, pierced through its entire length, and armed at one end with a clay tube or pipe. Everything being ready, fire was applied to the charcoal, and the whole mass instantly blown into a powerful heat through the reeds, the clay extremities of which were inserted in the basin', (Lock, op. cit., p. 122).

18   A similar case can be made for construction, which, until the development of prefabrication, involved, for the greatest part, the laying of stone upon stone, or brick upon brick, thus keeping the nature of a large part of its concrete labour constant for millennia. Unlike mining, however, construction has neither the problem of a production site dictated by nature, nor the problem of confined space.

19   'Modern mining methods use state of the art technology for drilling and blasting to reach the gold-bearing ores, which are then taken away for processing. The ancient Egyptians used slave labour. Agatharchides, a Greek writer from 150 BC visited some of the Egyptian mines and described

methods that had not changed for hundreds of years', Observer, 02.07.89.

20   Modern mining safety standards require at least two points of access to a mine.

21   By 'gold rush' we mean a frantic and uncontrolled scramble to extract gold from a known or suspected source regardless of the general condition of trading activity. This is not the same as 'gold quest', by which we understand the conscious searching for gold without any source necessarily suspected or identified, such search being driven by a sudden need for gold for the restoration or continuation of commercial intercourse. Some writers, e.g., Wachtel, H., 1990, loosely employ the former term to denote both conditions. This does not, of course, mean that these could not coincide, as they did, e.g., in the case of the Witwatersrand.

22   Apart from attempting to limit the size of claims, the dominion authorities in Canada in 1900 imposed the following charges in British Columbia: 'an individual miner's certificate (permit to mine) is $5; for a stock company of $100,000 capital or less, $50; for other companies, $100; for a Crown grant, $5; for recording a certificate or claim, or an affidavit, or generally speaking, any paper, the fee is $2.50; and when of unusual length, 30 cents per folio additional', Del Mar, op. cit., p. 428. Del Mar does not say whether and how the miners managed to pay this. When licence charges were imposed in Australia in 1854, insurrection broke out (see Buranelli, V., 1981, pp. 82-6).

23   This held right up till 1971, when the price of gold could once again proceed from its value.

24   Only the pen of a novelist can capture the richness/poverty of life/death which is a gold rush.

25   One of the major stumbling blocks to the development of a goldmining industry in the USA was its gold law, which allowed for the auriferous ore to be pursued underground beyond the boundary of the property upon which the works have been erected. This caused endless litigation and interruption of production, further adding to speculative pressures to make money by buying and selling shares to 'prospective' mines, rather than from the sale of production. (see, e.g., Lindley, C., 1897). On the Witwatersrand, the mines endeavoured to have all ambiguous property laws

clarified very early on, but also improved their position by buying up as much contiguous property as possible.

26 The value distinction between placer gold and lode gold is also borne out by technologically backward peoples tending to produce gold from the former, rather than the latter source, as Del Mar accounts:

> During the whole of the 17th century bands of predatory Cossacks roamed through these vast [Siberian] domains, prospecting, conquering, plundering the inhabitants, ransacking the Tschudi sepulchres, prospecting again, washing the river sands for gold, digging through the turf ...into the frozen placers below, or pointing out to the government where quartz mines might be located (for they themselves cares nothing for quartz), and where serfs or convicts might be led to delve for gold and silver (Del Mar, op. cit., p. 377).

Part One

TOWARDS A THEORY OF GOLD

Introduction

One of the most tenacious problems of economics, both Marxist and orthodox, is accounting for the behaviour of gold as an economic thing. Orthodox economists quickly exhaust their range of explanatory options when they either try to squeeze gold into their supply and demand curves, or crudely insist that things were better in the days of 'hard money'. Marxist economists either resort to simply quoting Marx (or other Marxists), or to dismissing the whole gold question with a few remarks.

No other commodity has displayed the kind of economic history that gold has. When something which has for millennia been synonymous with stability, invariability, constancy, etc., and suddenly, over a period of mere decades, comes to be associated with daily fluctuations, then there are some serious processes at work which require more than the kind of 'sectional interest' attention they have received to date. The nineteenth century *volte-face* in gold economics still awaits satisfactory explanation.

A theory of gold would have to distinguish between the material nature of gold, its production and its social form, and yet grasp the relationship between these. It must trace and account for the evolution of gold through its various social forms and lay bare the laws by which this evolution proceeds. It must trace its implications not only in the relation between gold and other things (objectified labour), i.e., in circulation, but also between gold and living labour, i.e., in its own production. In other words, a theory of gold must embrace a political economy of gold. These are the terms of reference of the first section of this work. It must be remembered throughout that we are dealing with a period when gold did actually circulate as money. Its conclusions cannot be transferred wholesale to the late twentieth century, since the social forms of gold which begin to emerge during the period under investigation and only reached maturity in the 1970s, particularly the doubling of its character as a commodity, are not examined here.

1 Gold as money

The exploration of gold specifically as money is necessary for three reasons: (i) the form on the one hand, has its own laws which are but tendencies in the commodity form and are also distinct from the laws of commodities, and on the other, demonstrates tendencies which become laws in the capital form; (ii) in the period under review, the dominant social form of gold was money, and not yet capital, although it was undergoing the transformation from the former to the latter during this period; and (iii) Marx's treatment of gold is incomplete in that his exploration of gold is ancillary to his exploration of capital.

We are looking at gold in a period when it made the transition from circulating as money, to: (i) its increasing confinement to hoards and (ii) its progressive marginalisation from the functions of money. In this dissertation, these developments are not followed through to the subordination of money economy to credit economy. Although silver also, for the greater part of this period, circulated as money, it is specifically excluded from this discussion. This is done partly because by the mid-1870s, silver had been almost completely eclipsed by gold as international medium of circulation and means of payment,[1] and partly because the relationship between gold and silver has no bearing on the exploration of the value of gold-producing labour-power.[2]

Gold and the functions of money

Gold has, until the twentieth century, always been associated with two attributes, viz., constancy and uniformity. These two attributes have consistently drawn comment in the literature, one scholar even making them the subject and title of a book (Jastram, R., 1977). Although this constancy and uniformity are each given proper attention and they are often discussed together, it is seldom that an inherent relation between them is brought out. Seldom, too, is a clear relation established between the *material* character of gold as a particular metal, and its *social* character as the money-commodity. While many scholars do point out that in all exchange societies the role of money eventually comes to settle upon the precious metals — and gold in particular — we could find no attempt to theorise why gold in one society appeared to command more or less the same value as gold in another, quite different and often quite distant, society. In other words, while much has been said about the role of gold both between and within societies, this distinction still warrants some theoretical attention. The suitability of gold to the role of money-commodity, though, has received much scholarly attention. Discussion, however, tends to be naturalistic. Gold is money because it offers certain natural qualities, rather than money is gold because it finds in gold certain natural qualities. An exception to this is Karl Marx, who will be discussed in a moment.

The naturalist approach begins by listing the attributes of gold: great density, homogeneity of substance, ease of division and reunification, malleability, longevity, virtual indestructibility, unadulterability, etc.,[3] as well as its widespread occurrence around the world. It is therefore 'logical' that it be used as money. The problem is that while these are, indeed, attributes of the metallic substance gold, they are adequate only to a *medium of circulation*, and not to money. Money is more than merely a medium of circulation. It is also a measure of value and a repository or store of value. This is generally recognised, but in a merely rhetorical way. Often it is simply asserted that money has three functions: a, b and c, these being simply quoted from Marx's *Capital*.[4]

The view most prevalent in the literature is that gold plays the role of the money-commodity because of the natural specification of gold. This view is not held in opposition to the view that gold is money-commodity because of the *social* specification of money, since the question of the social specification of money never occurs to most writers. It is not understood that, 'Though gold and silver are not by Nature money, money is by Nature gold and silver' (Gilliani, quoted in MEGA, Vol. II/9, pp. 77-78). Marx, and some other scholars, understand a particular form of social economy to posit money first, and then to seek out and settle upon that natural substance most adequate to money. The link between the attributes of the natural substance (gold) and the social substance (money) remains, in our view, only partially developed, even in Marx.

For Marx, exchange value posits its own independent bodily form out of exchange itself. This is so because exchange value stands as mediator between two distinct use-values, each addressing two distinct needs. Exchange finds a commodity the utility of which it will be to convey exchange value from one commodity owner to another. Exchange value itself is devoid of any material attributes, it is simply human labour in the abstract which has taken the form of abstract labour. It is, therefore, neither this, nor that particular labour, but yet capable of being any one of them. Labour objectified into a commodity is labour already performed. But such labour has not necessarily yet reached its final, desired objectification. Exchange value allows for labour objectified in one material form to be transformed into another, quite different material form. Thus, while it is but one exchange value, it is, at the same time, potentially all manner of use-values. It is potentially capable of satisfying all wants.

As long as exchange value must reside in some or other use-value, its identity coincides with the identity of that use-value. It is therefore as much subject to the attrition of nature as is the use-value which houses it and will decay along with it, unless transferred to another use-value beforehand. In other words, the labour objectified in a commodity runs the risk of not reaching its final, desired objectification on account of the transitory commodity decaying before it can be exchanged. Exchange value would therefore find itself best preserved in a commodity which is virtually indestructible. Gold is such a commodity.

Since the substance of exchange value is abstract labour, a homogeneous, undifferentiated substance, it can tolerate infinite division and re-unification *without ever ceasing to be abstract labour*. Exchange value, in order to be adequate to itself, eventually settles on a commodity which, no matter how divided or re-united, never ceases to be that commodity. It finds this attribute of itself most adequately reflected in the material gold. But this describes only the emergence of a *medium of circulation* out of exchange value. Measure of value and store of value remains unaccounted for.

The value of gold, like that of everything else, is determined by the amount of labour-power expended upon its production. In order to be universally capable of measuring the value of other commodities, its own value must be *universally uniform*. Exchange, after all, develops historically on the fringes of societies, i.e., it is an external, rather than an internal relation. For a particular commodity to be universally recognised as having a certain amount of value, it must be universally produced in more or less the same amount of labour-time. But societies offering gold to, and accepting gold from, one another have been extremely diverse, often standing at vastly different levels of social development. The average productivity of labour in those societies demonstrates as many different levels as their number. It would seem as though a measure could hold good only for one society, while every other needed its own measure. However, this was not the case.

Exchange value, which is the form in which objectified labour is carried from commodity owner A to commodity owner B, must persist through time from A to B. Time is therefore built into the very definition of exchange value. For a measure of value to have practical meaning also requires for its unit to be constant over time. A fluctuating measure of value interferes with the consummation of sale by purchase.[5] In other words, it undermines the *raisôn d'être* of exchange value. The tendency, therefore, is for exchange value to settle upon a commodity the value of which tends to remain constant over time. From the earliest times up till the mid-nineteenth century, the labour-process in gold production has remained constant. Marx does not develop this.

The explanation for the eventual emergence of a uniform measure for all societies lies outside the domain of gold as a natural substance. It might appear as if the homogeneity of its material would have a bearing on its role as *measure*, but it has not. Although, for all intents and purposes, one part of gold had as much value as any other, regardless of where each might have originated, the question is here not about the material of the ruler, but about its scale. Is the amount of value understood by society A to be embodied in one ounce of gold, the same as that amount so understood by society B?

Value expresses the fact that a commodity is a congelation of a certain *quantum* of labour, being a definite portion of society's general pool of labour. The exchange of commodities is, in effect, the exchange of labour. That such exchange is an exchange of equivalents presupposes that such labour is measured, which is the same as to say that it is reckoned in units. Each unit, if it is to measure all labour, must be of the simplest kind, so that all labour is only more complex than the unit by a certain degree, i.e., the unit must be recognisable in all labour. It must hence be capable of execution by the single individual unaided by mechanical devices or aided only by devices of such simplicity and crudity as to be the highest common denominator for all trading societies. This single unit of *labour* must, therefore, find expression in a single unit of a particular commodity. In other words, a single unit of a particular commodity must, *in its production*, objectify this single unit of labour. Gold, in its production, is such a commodity.

Wherever gold production has been indigenously developed in a society (whether as product, commodity or money), this initially entailed the exploitation of placer deposits including, occasionally, conglomerate outcrops.[6] Primitive placer gold production is differentiated from other gold production in that the gold was retrieved either by the unaided hand, or with the aid of the crudest and most minimal tools.[7] As gold occurs in nature in already metallic (native) form, no further labour needs to be expended upon it in order for it to become useful. The most basic form of concrete labour — simple, unaided labour — has been the character of gold-producing labour from the earliest human contact with it. The labour expended on gold production has thus immediately reflected the abstract unit of labour inherent in exchange value. In other words, a unit of abstract labour can, through gold, itself be given a material expression. The value of gold was, therefore, immediately identified with the labour-power that was expended on its production, rather than with the use-value of another commodity.

The widespread occurrence of placer gold throughout the world meant that in all societies, some sooner, others later, gold would not only come to serve as the measure of all value, but that the unit of that measure would be universally understood to mean the same thing: x amount of labour-time. The language of value was the first world language.

But this is thanks to an objective circumstance which has nothing to do with the natural qualities of gold and, rather, everything to do with natural occurrence. In as much as the very existence of gold is but an objective circumstance allowing the nature of exchange value to more adequately express itself, so is the nature of gold production an objective circumstance allowing for the emergence of a measure of value. While the placers lasted and gold lent itself to simple extraction, such simple extraction persisted, changing only rarely and in small measure. The form of gold production was therefore not only uniform across societies, but also constant over time. A measure of value persists only for as long as these two conditions are fulfilled. As *value*, gold is capable of preserving itself over time thanks to its indestructibility; as *exchange value*, its preservation depends on the constancy of the labour-process producing it. The importance of this distinction, which Marx does not make, emerges only when the process by which gold is produced begins to change.[8]

Gold and its method of production *together* offer to exchange value the means by which it might find a material embodiment most adequate to its inner character. It was therefore only a matter of time before exchange value would come to settle upon gold as the commodity which would liberate it from the stricture of use-values. 'Exchange value obtains a separate existence' (Marx, K., 1973, p. 145). Gold now appears as potentially all commodities.

As exchange value retains its quality of representing all use-values only as long as it remains exchange value, its inner tendency is to avoid commodities destined for consumption. It therefore eventually finds and holds onto a commodity least susceptible to physical consumption. It has already been pointed out that gold is, to all intents and purposes, indestructible. This means that it *cannot*, strictly speaking, be consumed. It is, however, a soft material and, as such, vulnerable to abrasion. It can be worn out of existence as a particular physical shape. But the particles to which it is reduced remain as much the material gold as was the larger lump whence they came. The original shape can again be reconstituted. Thus does gold escape the one axe which hangs over the heads of all commodities: consumption. Once produced, it exists forever. Exchange value settled into gold can thus ween itself from the mother which gave it birth: exchange. As value-for-itself, it seeks out and finds a commodity in which it can congeal as permanent value. Gold obliges it. Exchange value now exists for its own sake, and its role as an enabler of the socialisation of labour must now become secondary.

Wealth now becomes alienated from itself in that, far from willingly sacrificing itself on the alter of human pleasures, it must now, according to its concept, do everything to avoid such consumption. As value-for-itself its accumulation is not only obvious, but also measurable. It has become money. Money sets itself up as consumer of wealth in opposition to the human consumer, and in direct competition with him. Jacob draws a distinction between 'metallic wealth' and 'material wealth' (Jacob, W., 1968b, p. 115). The owner of wealth must now arbitrate between whether his person or his hoard will consume the gain. The hedonist and the miser are the pure forms of these extremes.[9]

Prior to the emergence of capital, or, more accurately put, where value is unable to assume the capital form, accumulation can take place only by the direct addition of one mass of values to another, thereby creating a new, larger mass. Accumulation arises from *exchange value*, and not from capital. Capital accumulation is but the perfection of this, requiring particular production and social conditions for its emergence. Many scholars, e.g., Hillel Ticktin, argue that by accumulation Marx means nothing more than accumulation of capital.[10] For this reason we let Marx speak for himself:

> Money, i.e., exchange value which has assumed an independent existence, is by nature the embodiment of abstract wealth but, on the other hand, any given sum of money is a quantitatively finite magnitude of value. The quantitative delimitation of exchange value conflicts with its qualitative universality, and the hoarder regards the limitation as a restriction, which in fact becomes also a qualitative restriction, i.e., the hoard is turned into a merely limited representation of material wealth. ...The degree in which the realisation of exchange value approaches such an infinite series, in other words how far it corresponds to the concept of exchange value, depends on its magnitude. ...But in passing one set of quantitative limits of the hoard new restrictions are set up, which in turn must be abolished. What appears as a restriction is not a particular limit of the hoard, but any limitation of it. The formation of hoards, therefore, has no intrinsic limits, no bounds in itself, but as an unending process, each particular result of which provides an impulse for a new beginning. Although the hoard can only be increased by being preserved, on the other hand, it can only be preserved by being increased.
> ...The activity which amasses hoards is, on the one hand, the withdrawal of money from circulation by constantly repeated sales, and, on the other, simple piling up, *accumulation*. It is indeed only in the sphere of simple circulation and specifically in the form of hoards, that the accumulation of wealth as such takes place, whereas the other so-called forms of accumulation... are quite improperly, and by analogy with simple accumulation of money, regarded as accumulation... Gold and silver constitute money not as the result of any activity of the person who accumulates them, but as crystals of the process of circulation which takes place without his assistance. He need do nothing but put them aside, piling one lot upon another, a completely senseless activity, which, if applied to any other commodity would result in its devaluation (Marx, 1977b, pp. 131-3. See also 1973, pp. 269-70).

But the accumulation of money for the sake of money is in fact the barbaric form of production for the sake of production, i.e., the development of the productive powers of social labour beyond the boundaries of customary requirements (Marx, 1977b, p. 134).

Because money is a more adequate form of expressing exchange value than is the commodity, it is therefore a more adequate means of exchange. It is better able to secure for its owner whatsoever use-value he may require, unlike exchange value locked into a particular use-value. Money therefore presents itself as a more adequate form in which to hold value, and a more adequate form in which to accumulate.

Here two most important transformations take place. Firstly, accumulation, which is the movement of 'the contradiction between free time and labour time' (Marx, 1973, p. 711), and which hitherto became active only in the actual exercise of labour-power itself, now assumes corporeality. The contradiction between free time and labour-time comes to reside in a single commodity: money. Money is the physical, material, bodily, expression of this contradiction since it is *at one and the same time* both exchange value in general and the particular exchange value which is its bodily form. As exchange value in general, it is 'free time' in that it is potentially *all* use-values, both in their variety and in their quantity; as particular exchange value (money is always just so much money, a given amount) it is a given quantity of use-values. It is capable of exhaustion and thus presents itself as a given quantity of social labour-time. The contradiction between free time and labour-time here takes the physical form of money, and as money, expresses itself as a contradiction between its physical or quantitative limitation, and its essential limitlessness. It is simultaneously exchangeable against all exchange values (in its social aspect) and exchangeable against only a fixed amount of exchange values (in its concrete aspect). It is therefore *inherent* in money to seek to overcome its own physical limitedness in order to render it adequate to its own essential limitlessness. This it does by constantly seeking to expand itself.

In this way, while the *product* is an objectification of concrete labour, *money* is an objectification of social labour. The movement inherent in the exercise of labour-power to constantly seek to expand its free component at the expense of its necessary component (thereby accumulating free time) now re-emerges in money as a movement to transform money into more money, thereby accumulating the objects of free time. Money is therefore the first objectification of labour in all its aspects. Indeed, money can, more adequately than any other form of product, be described as 'a form of labour', since in money, like in labour-power, is its accumulation driven by itself.

> We have already seen, in the case of money, how value, having become independent as such—or the general form of wealth—is capable of no other motion than a quantitative one; to increase itself. It is according to its concept the quintessence of all use-values; but since it is always only a definite amount of money (here, capital), its quantitative limit is in contradiction with its quality. It is therefore inherent in its nature constantly to drive beyond its own barrier. ...Already for that reason, value which insists on itself as value preserves itself through increase; and it preserves itself precisely only by constantly driving beyond its quantitative barrier, which contradicts its character as form, its inner generality. Thus growing wealthy is an end in itself. The goal determining activity of capital can only be that of growing wealthier, i.e., of magnification, of increasing itself. A specific sum of money (and money always exists for its owner in a specific quantity, always a specific sum of money) ...can entirely suffice for a specific consumption, in which it ceases to be money. But as a representative of general wealth, it cannot do so. As a quantitatively specific sum, a limited sum, it is only a limited representative of general wealth, which goes as far, and no further than, its exchange value, and is precisely measured in it. It thus does not by any means have the capacity of buying all pleasures, all commodities, the totality of the material substances of wealth; [First is the capacity of the commodity to function effectively as exchange value limited by its specific use-value, then, once posited as money, is its capacity to function effectively as this new use-value limited by its specific exchange value, FS] ...Fixed as wealth, as the general form of wealth, as value which counts as value, it is therefore the constant drive to go beyond its quantitative limit: an endless process. Its own animation consists exclusively in that; it *preserves* itself precisely as a self-validated exchange value distinct from use value only by *constantly multiplying* itself (ibid., p. 270, emph. orig.).

Historically this takes the form, amongst all peoples in the early stages of their trading development, of money stumbling its way from commodity to commodity, more or less by trail and error, each commodity in turn taking up the role of money. 'The particular use-value which, as a result of barter between different communities, become commodities, e.g., slaves, cattle, metals, usually serve only as the first money within these communities' (Marx, 1977b, p. 50). Since there are definite economic specifications which the money commodity has to meet, the different commodities selected at different points to play the role of money are variously suited to this function: (i) its labour content must be appropriate to the range of magnitudes of transactions, i.e., to the amount of labour changing hands;[11] (ii) once acquired, it must preserve itself long enough to be able to exchange without loss for the desired use-value, when such comes along; (iii) it must be available in sufficient discreet units in order that circulation (its velocity notwithstanding) may not be interrupted for want thereof.[12]

With this kind of specification it would be only a matter of time before this role would devolve to the metals and among them, to the noble ones. Hence, eventually, silver and gold (particularly the latter) coming to be universally used as the money-commodity by communities separated by both space and time. Of all known materials, gold is by far the most suited to the role of money.[13] 'Though gold and silver are not by Nature money, money is by Nature gold and silver' (loc. cit., p. 25).

When exchange value has 'taken over' the physical body of another commodity as its own particular materiality, it has done two things: Firstly, it has elevated itself above the world of commodities; and in so doing it has, secondly, elevated the commodity thus taken over so that it, too, now appears to be elevated above the world of commodities. From this point on, gold becomes a mystery.

The commodity which is also the money-commodity comes to be produced for its own sake. Its production ceases to relate to its consumption, or, more precisely, its production as the money-commodity is unrelated to its consumption as a commodity. The money commodity is the first commodity produced *for exchange*, rather than for consumption. Unlike capitalist commodity production, where production is for exchange, but is nevertheless conditioned by consumption, no such condition restrains the production of the money commodity. Quite regardless of their concrete reality commodities are as products, produced in order that they may be consumed, even if after a lengthy delay. Indeed, they are exchange values for this very reason. As values, they cannot be value-for-itself since they are inextricably tied to use-values, which, if not sooner or later consumed, will cease to be both use-values *and* values. Value-for-itself needs to be free from the threat of eventual consumption. Once money posits value-for-itself, however, then the money commodity ceases to be a commodity for consumption (according to the concept of commodities).[14] This commodity is now specifically produced for accumulation.

Hence intensified production of that commodity which is set aside to play the role of money — strictly speaking, this is, in fact, intensified *appropriation* in that it applies as much to plunder, free labour, etc., as to forms of production externally akin to slavery. That this applies to the *money*-commodity is why it is also evidenced in the case of silver, hence, e.g., Potosi. Silver never occurs as placers. Placers allowed intensified production to take different forms because diffused occurrence allows uncontrollable access to the *material* of production, i.e., it could not support slave-like production, the exception to this being where a sufficiently large coercive force can be mustered to control a work force scattered over a large, borderless area, or where a region is so hostile as to guarantee death after escape.[15] Hence continued *appearance* of placer gold production as if it were the production of a *non*-money-commodity. Placer mining would only begin to assume character of money-commodity production once the placers begin to run out.

Gold and its 'constant' value

Value preserves itself in use-value, the only form in which equivalence can be expressed. Any deterioration in that use-value, i.e., in its capacity to satisfy the need associated with it, implies a decline in its capacity to officiate as equivalent. In other words, a decay in its use-value results in a decline of its own relative value. But separately, commodities are also subject to a continual decline in their value as the level of productivity with which they are produced increases. Gold is no exception. When dealing with the products of extractive industry, such as gold, a crucial qualification to this statement has to be made. This is done when we return to this point, below. A distinction must at this point be made between a decline in value on account of *a deterioration of use-value*, and a decline in value on account of *an improvement in the productivity of labour*.

In the case of gold this distinction is most crucial, for the metal presents a double deception: (i) on account of its virtual indestructibility, the value of gold cannot decline as a result of a decay of its use-value, since the latter preserves itself; (ii) on account of the actual production process of gold which for centuries has lent itself only with the greatest tardiness to improvements in productivity (See Jacob, W., 1968a, Chs. II, VII and X). Reminding ourselves of the discussion in Part III of the Introduction, in the case of placer production, productivity was not improved, because production was too easy; in the case of underground mining, because to do so was too difficult. The value of gold has, as a result, declined but slowly, remaining constant for very long periods of time. Fluctuations in the value of gold tended to come about by its gradual rise as a result of the gradual depletion of sources, and its sudden fall occasioned by plunder or the discovery of new sources. These *objective* circumstances combine in their effects to give gold the appearance of being above the law (of value), so to speak. Many mystical accounts of the 'power' of gold have their bases in this. Even Jastram's statistical odyssey, *The Golden Constant*, is, ultimately, reducible to this (Jastram, R., 1977).

More than this, given that the value of gold, for these purely accidental reasons, remains relatively constant, or has historically diminished only very gradually, it allows an accumulation of value simply on the basis of an adjustment in the value ratios, which is almost always in favour of gold.[16] For a long time, therefore, the owner of gold could accumulate value by doing nothing at all. The amount of 'human pleasures' which a given quantity of gold could command would increase purely on account of a diminution of the values of these 'pleasures' standing opposite it. Here we speak of gold not as money-commodity, but simply as commodity, i.e., not as an enabler of exchange, but as one of the parties to it.[17] Gold did not only preserve wealth, it appeared to expand wealth without itself actually expanding. The certainty of value pursuing its own self-expansion appears here to be identical with gold *as a physical thing*. Any diminution in the value of gold would therefore deal a double blow to the sanctity of wealth held in that form, and would impel the development of the other forms of wealth preservation.

Decline in the value of a commodity announces itself to its owner first in the deterioration of its use-value while still in his possession and only in the second instance in a decline in its rate of exchange against other commodities, i.e., upon its alienation. It therefore seems, to the owner of commodities, as if the value of an item is suggested by the condition of the use-value and that this is merely confirmed by its exchange value. From this vantage point, socially necessary labour-time itself appears to be an irrelevancy which has nothing to do with value, and thus an illusion.[18]

If this is true for commodities in general, then it is even more true for gold in particular. Not only does the use-value of gold not deteriorate, the preservation of its value which appears to its owner to flow from this, has for centuries seemed confirmed by the actual constancy of value (reflecting the unaltering productivity of its production process). Wealth can, it would seem, without fear of diminution, be amassed in this form. When use-value and value are conflated, the illusion that the value of gold remains constant is reinforced.

Gold therefore appears as the ideal incarnation of value-for-itself, most eminently suited to the task of accumulation. While accumulation is a process of accrual, time, as described above, imposes a deduction upon that accrual. With gold this deduction appears to have been suspended. Accumulating wealth in the form of a commodity other than gold is like accumulating water in a leaky bucket. When that commodity is gold, the bucket appears leak proof. This phenomenon must, at least in some way, contribute to the mystery surrounding this material.

But objective circumstances are precisely that, and circumstances change. A major change in circumstances came with the flooding of Europe with gold from the New World during the sixteenth century.[19] Almost overnight (compared to centuries of constancy or only very gradual rise) the value of gold declined very rapidly. All of the new gold entering the world economy from the New World did so by way of a method far more productive than mining: plunder.[20] A new social average had to emerge, taking account of tonne per *Conquistador* per annum. The result was that all existing hoards suffered loss (including those in the holds of the arriving ships, for their value was anticipated at the prevailing world average). The value of gold had to decline, no matter how well it had preserved its use-value. One unit of gold could henceforth buy less. In other words, there was a universal increase in prices.

This has nothing to do with the *quantity* of gold coming into circulation, as is so often suggested, especially by theorists of the sixteenth century price revolution, and latter-day gold standard protagonists. The debates surrounding the quantity of specie in circulation in relation to the level and total of prices does not concern us here. We will just briefly say that a change in the quantity of gold in circulation, all else remaining constant (i.a., the values of gold and commodities, the sum of prices, the velocity of circulation), results only in increased hoarding or lending (or other forms of advancement). The same quantity still expresses the same sum of prices as the value stays the same.

Change in the value of gold, however, alters the ratio in which it expresses the prices of commodities. An increase in value means that the same quantity of gold now expresses more value than that contained in the sum of commodities. Commodities, provided that their values have not altered, now have their values expressed in smaller amounts of gold. In other words, there has been a universal drop in price. Likewise does a decline in the value of gold reflect in a universal rise in price. The gold (and silver) from the New World that flooded Europe in the sixteenth century did, of course, increase the quantity of gold in circulation, but to theorise the concomitant universal increase in prices simply in terms of supply and demand is to miss the fact that a value revolution had taken place. New World treasure was procured by plunder, a very much more efficient form of production, thereby setting a new average necessary labour-time for gold production. No similar revolution occurred in the values of commodities at the time, hence the higher prices.[21]

The commodity which serves for the universal expression of prices does for other commodities what they cannot do for themselves. It expresses a particular commodity's *value* alongside the values of all other particular commodities in a unified series centred on its own use-value and called *price*. In other words it, and only it, has a utility which springs from the exchange of commodities itself. It is necessary now to examine this utility in contradistinction to its 'natural' utility as commodity.

The two use-values of the money-commodity

There are two dimensions to the *social* character of gold: it is, first and foremost, a commodity, i.e., a use-value in the social form of a value. It is therefore, like all commodities, subject to the laws of commodities; but in addition to this, it is also money, and hence subject to the laws of money. Indeed, it is only as a commodity that it can become money: the one commodity set aside from all others so that they may all use *its* use-value in order to express their own value for one another. Gold is therefore the medium through which all other commodities express their values relative to one another, forfeiting the relative expression of its own value.[22] Marx encapsulates it thus:

> In so far as the price of a commodity is *realised* in gold, the commodity is exchanged for gold as a commodity, as a particular materialisation of labour-time; but in so far as it is the *price* of the commodity that is realised in gold, the commodity is exchanged for gold as money, i.e., for gold as the materialisation of general labour-time (Marx, 1977b, p. 91).

In addition to its consumption as particular use-value (tooth filling, jewellery, etc.), gold is now allocated a consumption by its social role as abstract wealth. Every product is, in its concreteness, a use-value. This it is regardless of the social form under which it might have been produced. The social form based on commodity exchange lends all use-values the additional character of exchange values. As particular use-value, gold, too, acquires an exchange value. Hence are jewellery, tooth fillings, electronic circuitry, cladding, etc., all use-values as well as exchange values. However, having once posited the money form, exchange imbues the commodity to which the role of money falls, in this case gold, with an *additional* use-value. And as this second use-value it does *not* posit exchange value, but is, nevertheless, ascribed one, as we shall see again later.

> While commodities thus assume a dual form in order to represent exchange value for one another, the commodity which has been set apart as universal equivalent acquires a dual use-value. In addition to its *particular use-value* as an individual commodity it acquires a *universal use-value* (ibid., p. 47, emph. ours).

This 'universal use-value', of which Marx speaks, is a use-value associated with the specific needs of no particular individual. It is associated with a need of this *form of production* as such. It, 'satisfies a universal need arising from the exchange process itself, and has the same use-value for everybody—that of being a carrier of exchange value or a universal medium of exchange' (ibid., p. 48).

Generally, in a discussion of use-value, the term 'use-value' would ordinarily suffice without further qualification. In a discussion of the money commodity *per se*, however, the term becomes too broad since we now have to distinguish between particular use-value and universal use-value. We wish to propose a number of categories in terms of which use-value — and in particular the use-value of the money commodity — might be further explored. The distinction already made by Marx above will serve as our starting point. For 'particular use-value' (shoes, road, jazz, tooth filling, etc.) we propose 'use-value I' or 'private use-value'; for 'universal use-value' (money) we propose 'use-value II' or 'social use-value'. Throughout the remainder of this document, the terms 'use-value I' and 'use-value II' will be used, and shall to be understood to mean, 'particular use-value' and 'universal use-value' respectively.[23]

Since the money form of value is a higher form than the commodity form, use-value I is subordinated to use-value II (in the same way as, e.g., money is subordinated to capital), or, in different words, this second use-value, this universal use-value, is posited out of circulation and as such, therefore, *presupposes* exchange value. It is therefore a *more developed form* of the product of labour than is a commodity. Use-value II now posits use-value I as a less-than-ideal form of itself, as potential use-value II. Thus does it establish itself in the popular psyche as something more valuable than a common article of utility. It therefore becomes 'wasteful' to use gold for anything other than money, except where such use serves as hoard, or in the narrow function of a deliberate signal of social station.

> Gold and silver articles, quite regardless of their aesthetic properties, can be turned into money, since the material of which they consist is the material of money, just as gold coins and gold bars can be transformed into such articles. Since gold and silver are the materials of abstract wealth, their employment as concrete use-values is the most striking manifestation of wealth (ibid., p. 134).

Gold does not cease to be jewellery, tooth filling, cladding, etc., just because it has become money. And since: (i) it is not jewellery, tooth filling or cladding as such which becomes money, but the *material* out of which these have been fashioned; (ii) the material can be subdivided and re-amalgamated at will; and (iii) it cannot be consumed in the sense of other use-values, adapting its shape to the demands of use-value II is as easy as adapting it to the various demands of use-value I. Whether it changes from gold leaf to wedding ring to tooth filling involves the same process as a change from any of these into coin, medallion or bullion.

Because one is dealing here with the material as such, one can abstract from the varying quantities of additional labour required to bring about these changes in shape. A great deal of value, e.g., is added in smelting, crafting jewellery and making chains, the latter two of which combine a high degree of craft skills with a high degree of mechanisation.[24] Since it is as a *material* that it figures as use-value II, the socially necessary labour-time for its production is that which is required to retrieve the material from nature and not that necessary to give it this or that shape. It is already in an adequate shape *as it is found* without any further labour having to be expended on it. All that is reckoned with is the labour expended in producing it (which includes the labour of transporting it to the point where it may enter circulation).[25] It can therefore quite easily serve the demands of use-value II while retaining a particular shape associated with use-value I though not necessarily in the most convenient way. The Nazis did not expect the Jews to first melt down and mint their own gold teeth before they could recognise them as money, neither did Mussollini complain about his getting the funding for his Abbysinia campaign in the form of Italian housewives' wedding rings (Green, T., 1968, p. 12). Indian peasants cover their womenfolk with gold jewellery at harvest time as a form of saving for leaner times (Green, 1987, pp. 135-7).

Use-value I therefore serves as a huge decentralised hoard for use-value II (in the same way as, at a higher level, the money form serves as hoard for the capital form). For the demands of use-value II, even though discreet and equal units of weight is most convenient for its day-to-day purposes, for its essential purpose even the raw natural state of gold as nuggets, grains, dust or flakes will suffice. What is important is the *material* and not the shape in which it comes.

The point is that in practice it requires no more and no less labour for this material to be ensconced in the role of use-value II than it does for use-value I, though, in essence it actually requires less. Although in practice the adaptation of gold from a use-value I shape into one more suited to use-value II (the minting of coin and the casting of bullion bars) involves an additional labour expenditure, there is no *essential* reason why such additional expenditure should have to be made. Indeed, as is implied in these lines, it is as use-value II that the labour expended on the production of the material *per se* is most truthfully revealed.

Marx gives only slight attention to the money-commodity in its aspect as commodity, and, indeed, appears somewhat dismissive of it. He says that, 'just as the precious metals are useless in the direct process of production, so they appear to be unnecessary as means of subsistence, i.e., as articles of consumption. ...Their aesthetic qualities make them the natural material of pomp, ornament, glamour, the requirements of festive occasions, in short, the positive expression of supra-abundance and wealth' (Marx, 1977b, pp. 154-5). Here Marx appears to forget two of his own lessons: firstly, 'Money—...i.e., the universal commodity—must itself exist as a *particular* commodity alongside the others, since what is required is not only that they can be measured against it in the head, but that they can be changed and exchanged for it in the actual exchange process' (Marx, 1973, p. 165, emph. orig.);[26] and, secondly, that a commodity is 'an object of human wants, a means of existence in the widest sense of the term' (Marx, 1977b, p. 27), or, as later put in *Capital*, 'a thing that by its properties satisfies human wants of some sort or another. The nature of such wants, whether for instance, they spring from the stomach or from fancy, makes no difference' (Marx, 1983, p. 43).[27]

Subsistence, which means 'existence in the widest sense of the term', embracing the satisfaction of all 'human wants', whether 'they spring from the stomach or from fancy', is here reduced to strictly those wants which spring from the stomach. Fancy, as a fountainhead of wants, is here dismissed. He is now saying that the commodity which becomes the money-commodity is not really a commodity in the first place,[28] so that, 'any quantity of them can thus be placed at will within the social process of circulation without impairing production and consumption as such' (Marx, 1977b, p. 154). But he goes further, suggesting that whatever use-value gold (and silver) might have, such is confined to 'the positive expression of supra-abundance and wealth'. If something becomes a use-value only because its utility is to express wealth, then how could it have been a commodity to be set aside as the money-commodity in the first place? If its only use-value is to express wealth, then it is *as use-value* already the product of a social relation. This makes gold *in its physical concreteness* an abstract thing. This apparent laxity on the part of so rigorous a scholar again suggests to us that Marx has not made a study of the political economy of gold as such.

It is important that the distinction between use-value I and use-value II be kept clearly in view, in order to know whether the particular phenomenon, tendency or law being observed stems from one or the other aspect. The ability to explain the observation will depend on this.

Notes

1. Large silver-based economies persisted though, e.g., India, China, Spain.
2. We offer a brief mention of the silver:gold value-ratio question in Section II.
3. 'Neither rust nor worms can spoil this metal', according to Sappho.
4. See, e.g., Innes, D., 1981, for a particularly mechanical study in which, money 'in order to function as money... must of necessity adopt the commodity-form', p. 6.
5. With the transformation of gold from money into capital, this interference becomes an opportunity for accumulation, a question outside the scope of this thesis.
6. One prospector reports that in the eastern Transvaal, 'a body of Kafirs were at work breaking out quartz, in which the free gold was easily visible' (Lock, op. cit., p. 20).
7. Chapter I and Section II of this dissertation offer a fuller account of this.
8. This is taken up in Section II.
9. Die Genußsucht in ihrer allgemeinen Form und der Geiz sind die zwei besondern Formen der Geldgier. Abstrakte Genußsucht unterstellt einen Gegenstand, der die Möglichkeit aller Genüsse enthielte. Die abstrakte Genußsucht verwirklicht das Geld in der Bestimmung, worin es der *materielle Repräsentant des Reichtums* ist; den Geiz [verwirklicht das Geld — FS], in sofern es nur die allgemeine Form des Reichtums gegenüber den Waaren als seinen besondren Substanzen ist. Um es als solches zu halten, muß er alle Beziehung auf die Gegenstände der besondren Bedurfnisse opfern, entsagen, um das Bedurfnis der als solcher zu befriedigen (MEGA, Vol. II/1.1, p. 147).
10. Personal communication, January 1992.
11. 'Copper was adopted as the basis of money in Sweden in 1625. ...Worth only one-hundredth the value of silver it was unsatisfactory for ordinary payments because of their great weight. While burglars could not steal the money because their could not carry it, wagons were needed for ordinary payments' (Kindleberger, C., 1987, p. 23). If gold were common coinage today, it would present the same prblems.
12. Silver mining was disrupted by the Black Death after 1348, resulting in an acute shortage of money during the expansion

of trade following in its wake. Barter expanded for want of an adequate money-commodity. Salt and pepper, especially, stepped in as equivalents (ibid., p. 24).

13 In the historical sequence of metals to play the role of the money commodity, the next in line after copper, silver and gold will have been platinum. Although some platinum is being hoarded in bullion form alongside gold as a 'store of value', its emergence to prominence as a monetary metal has been eclipsed by the development of the credit system to overall world dominance. The study of this development is not part of this thesis. In the meantime, it is unlikely that anyone will ever be praised for having a heart of platinum.

14 In practice, though, this often devolves onto a commodity for consumption the longevity of which extends beyond those of commodities around it. In such cases the money commodity stands to be replaced by something more immortal, if and when such should come along. An example might be many peoples in Africa for whom cattle originally served as money, this later to be replaced by coins as their economies merged with those using money in this more ideal form.

15 Hence the success of the Russian state in working the great Siberian placers with serfs and convicts. For descriptions of Russian goldmining in the nineteenth century, see Del Mar, op. cit., pp. 372-96 and Lock, op. cit., pp. 369-455. Similar control provided by the impenetrable Amazon allowed for placer gold extraction on the basis of unfree labour:

> If the whole number of captives and slaves forced to work in the gold mines of Brazil be multiplied by the years they worked, and the quotient divided into the entire product of the mines, it will be found that they produced on the average less than $40 a year per man. Of this amount one fifth went to the Crown at the outset, leaving only $32 per man. By the time the mine proprietor turned this into coins it would be reduced to about $26, thus reducing the average product to about half an American gold dollar per week per man, less than the solitary gleaner still wins in freedom from the neglected corners of the abandoned catas. The imagination can scarcely realize the cruelty and privation to which the mining slaves were formerly subjected in order to 'make them

pay' the cost of their maintenance (Del Mar, op. cit., p. 253).

Free mining was attempted in Siberia, especially where operations have been abandoned by the state. This generally failed unless either it was adapted into a privately owned venture worked on the basis of unfree labour, or, as after the mid-1860s, technological improvements were introduced (ibid., pp. 385-6, and Lock, op. cit., pp. 448-55).

16 Given the general backwardness of economy prior to the advent of surplus-value production, value in general diminished only very gradually.

17 Jacob offers an idea of 'the high value of money, or the low cost of all the necessaries and luxuries' in 1189, by a lengthy examination of the coronation banquet expenses of Richard I (see Jacob, W., 1968a, pp. 329-37).

18 Many an economic head has come unstuck over this illusion. Alec Nove, prolific writer on 'market-socialism', for example, talks of Marx's, 'artificial and unjustified separation between value and use-values', arguing that 'some things have more use-value than others', and must therefore command a higher price (Nove, A., 1983, p. 21).

19 For the much-explored subject of the effects of New World gold on the economy of Europe, see, e.g., Jacob, W., 1968b, Del Mar, op. cit., Vilar, P., 1976, Supple, B., 1964, etc. Ramsey, P., 1972 suggests that these effects tend to be exaggerated in the literature. For a summary of the monetary theories that arose in the wake of these developments, see Vickers, D., 1968, and also Rubin, I., 1979.

20 Even if this gold were originally produced as exchange value, which it was not, its existence as exchange value would have abruptly ceased with its plunder, regardless of how efficiently or otherwise it may have been produced.

21 'The sudden and forcible transfer of hoarded money from one country to another is a specific feature of the ancient world; but the temporary lowering of the production costs of precious metals achieved in a particular country by the simple method of plunder does not affect the inherent laws of monetary circulation', Marx, K., 1977b, p. 161. For Marx' critique of David Hume's quantity theory of money, see ibid. pp. 160-65 and Marx, 1983, pp. 118-9. For Ricardo's role

in the 'bullion debates', see Gordon, B., 1976, and also Marx' critique of this role in Marx, 1977b, pp. 174-9.

22   'We overlook the facts that gold, when a mere commodity, is not money, and that when other commodities express their prices in gold, this gold is but the money-form of those commodities themselves' (MEGA, Vol. II/9, p. 90).

23   Under use-value I a distinction can, of course, be made between use-values individually consumed (shoes, jazz, tooth filling) and those collectively consumed (road, jazz, weather report) — hence individual use-value and collective use-value. No such distinction is possible under use-value II, since its consumption is entirely collective. Even the term 'consumption' requires some qualification when use-value II is gold, as has been suggested.

24   For the suitability of gold to goldsmithery, see Wise, E. (ed.), 1964, pp. 253-9. For the intricacies of gold chainmaking, see Green, T., 1987, pp. 157-9.

25   For more on seignorage, see Kindleberger, op. cit., pp. 30-1.

26   'The commodity which has been set apart as universal equivalent acquires a dual use-value. In addition to its particular use-value as *an individual commodity* it acquires a universal use-value', Marx, K., 1977b, p. 47, emph. ours. Or even more emphatically: 'The use-value of the money-commodity becomes two-fold. In addition to its special use-value as a commodity (gold, for instance, serving to stop teeth, to form the raw material for articles of luxury, &c.), it acquires a formal use-value, originating in its specific social function', 1983, p. 93.

27   'Neither are we here concerned to know how the object satisfies these wants, whether directly as means of subsistence, or indirectly as means of production', ibid.

28   'Herodotus ...says, "the Macrobian Indians, ...took them to a prison for men, where all were confined in golden shackles; brass, with these Ethiopians, being the scarcest and most esteemed of the metals"'(quoted in Lock, op. cit., p. 4). There were, though, cases where gold, though not produced as money, was produced for exchange, despite its having no particular use-value in the society producing it. Thus, does Lock report that, 'During his [Mungo Park's] stay in Kamalia [pre-colonial West Africa], the gold collected by the traders, for salt alone, was nearly £198 sterling' (ibid., p. 36).

2 The money-commodity and the value of labour-power

As has been suggested in the introduction to this section, the interaction between gold and the economy in general does not only take place where it has to circulate other commodities, but also where it is itself an object of circulation in its own right. This occurs in its production, where it exchanges with labour-power. Marx does not explore the production of the money-commodity with the same thoroughness as he does its circulation. His mention of over-work and barter at the site of production are but footnotes to his main argument, which, in the case of the money-commodity, actually relates to circulation. His discussion of the production of exchange value as such remains sketchy and over-reliant on the dramatic impact of Siculus' descriptions of the horrors of silver mining. The production of exchange value must be examined both on the basis of labour *not* as itself exchange value, and on the basis of labour *as* itself exchange value.

The value of labour-power in the production of money

As explained above, production of the money-commodity is driven not by consumption, but by exchange economy itself. The commodity which gives concrete form to abstract wealth is produced, seemingly, for its own sake. The dynamic behind this has been explored above only in so far as it affects consumption. What interests us now is how a relation emerging in exchange throws its reflection back into production, there to do to labour in production what it has already done to labour in consumption. But the most serious constraint on the value of labour-power of the goldminer is provided quite independently of any variation in the values of gold, commodities, or commodity-producing labour-power. It is provided neither by gold as use-value I, nor by the technical aspects of mining, but by the *social* nature of gold, gold as use-value II.

The tension set up in circulation between whether abstract wealth is consumed or preserved, in production takes the form of a tension between necessary labour-time and surplus labour-time.[1] At bottom, this remains a curtailment of **consumption** — that consumption which takes place *within* the exercise of **labour-power**. The cost of labour-power, whether a value or a product, is a direct deduction from the total product, i.e., a direct deduction from accumulation. The two portions of labour-time enter into conflict with each other only when the product assumes the form of the material embodiment of wealth in general. Each portion is here measured in terms of wealth in general, and not in terms of the consumptive requirements of the gold producer. The gold producer himself, even if he is a free worker — indeed, especially if he is a free worker — sees the value of his own labour-power as curtailing the ability of the money he digs out of the ground to accumulate. His own need to reproduce himself confronts him as destructive of his wealth. His affirmation as human being seems to him his denial, his denial an affirmation. This is the most elementary moment of the contradiction between living labour and objectified labour. It is to become more developed and generalised with the generalisation of wage-labour.

The gold worker gives expression to the contradiction between consumption and accumulation by denying to his own labour-power that which is necessary for its reproduction. Necessary labour-time is kept to an absolute minimum and that minimum incessantly pushed back in order that surplus labour-time might be extended. Value of labour-power in goldmining, $v_{LP(G)}$, is constantly reigned in, in order that surplus-value, s, might be increased. The application of technology to mining is prompted by a different set of movements and must here be left out of the discussion (a discussion of this is offered in Part II). Necessary labour-time in gold production can, consequently, only be reduced by reducing $v_{LP(G)}$. This is achieved by undermining the standard of living.

The private one-man gold-digger (the free miner) must become contemptuous of his own life as he tries to reduce the value of his own labour-power in order that he might accumulate his own surplus-value *as money*.[2] The near-bestial existence of the digger is well known, and has always been the subject of novels and the cinema — hilarious in its brazen disregard for the niceties of civil (and civilised) society. In so far as the advancement of society consists in the material and cultural advancement of its constituents, a society of private individual gold miners can only be a regressive one. Illustrations abound from around the globe. At the mines in Paramillo in Argentina, e.g.:

> some 4 or 5 miners were working in 3 of these mines, but only extracting fragments of ore from old shafts and levels; these ores hold 7 to 10 oz. of fine gold per ton, but the quantity is so insignificant, that it can only afford a miserable existence to such men (Lock, op. cit., p. 200).

Where gold is produced *for others*, capitalist forms excepted, the reduction of the value of labour-power does not end when the worker has reached 'a miserable existence', but proceeds all the way to his death. The expropriator of the surplus-value of the gold producer turns virtually the entire product into surplus-value.

> In any given economic formation of society, where not the exchange value, but the use-value of the product predominates, surplus-value will be limited by a given set of wants which may be greater or less, and that here no boundless thirst for surplus-labour arises from the nature of the production itself. Hence in antiquity over-work becomes horrible only when the object is to obtain exchange value in its specific independent money-form; in the production of gold and silver. Compulsory working to death is here the recognised form of over-work. Only read Diodorus Siculus (Marx, 1983, p. 226).

Throughout history, such labour has always been characterised as 'slavery'. One reads of 'gold slaves', 'goldmining slavery', etc. The accuracy of such a characterisation extends, however, only as far as its external appearance. Brutal and inhuman as it might have been, slavery, as a form of economy, allowed for the reproduction of the labour-power of the slave, albeit at a very primitive level. Producing gold as money *for others* definitely proceeded on the basis of the physical destruction of the worker, his working to death.[3]

To briefly anticipate an argument which properly belongs to a later point in this discussion (see Parts II and III, below), it may be pointed out that goldmining capital here appears to be more humane than the self-employed gold digger. The goldmining capitalist will make a strict economic calculation of the volume of investment required against the expected returns, knowing that there are distinct limits to the extent to which the value of labour-power can be reduced. He operates in a world of exchange of equivalents. The private one-man gold digger, though, does not go on strike against himself. He does not break any minimum wage laws. He sacrifices the value of his own labour-power in the hope of turning his non-wealth into wealth.

The effects of different value changes on the production of gold

Although Marx does distinguish between the value of gold as its own labour value and the value of gold in terms of other commodities, this distinction is only conceptual. He does not actually investigate empirically when a change in the value of gold is on account of an actual change in the labour-time required for its production, and when it is on account of a relative change in relation to commodities. That the cause of change in the value of gold itself is subject to change therefore escapes him. He does not anticipate the setting of gold production on a new basis, entirely new value relation, although he does anticipate its eventual depletion as a resource.[4] This distinction forms an important part of the remainder of this discussion.

We will first consider the relations between the values of gold-producing labour-power, $V_{LP(G)}$, gold, V_G, commodities, V_C, and commodity-producing labour, $V_{LP(C)}$, as simple relations between commodities, with gold serving only as measure of value and medium of circulation, i.e., as relations of simple circulation. Separately from this we will then consider these same relations as determined by money. There are hence two complexes of relations which will be explored: on the one hand, the relations between $V_{LP(G)}$, V_G, V_C, and $V_{LP(C)}$ with money posited by commodities (C - M - C); and, on the other hand, these same relations with commodities posited by money (M - C - M).

A distinction has to be made between the *substance* of value and its *expression*. The former is the amount of labour-time recognised by society as necessary, on the average, for the production of a particular use-value; while the second instantiates that amount in terms of units of other use-values. The former expresses a relation of economy in general; while the latter expresses a relation of commodity-economy in particular. Because the material gold came universally to play the role of the use-value in terms of which the values of other commodities are expressed, this distinction is important if a political economy of gold is to be constructed. It must be clear that the *expression* of value presupposes its substance.

If the values of commodities are expressed in terms of gold, changes in the values of commodities, v_c, register as changes in the quantities of gold that they will command as equivalents. It follows, equally, that changes in the value of gold, v_G, will be reflected in changes in the quantities in which various commodities will command fixed amounts of gold. But this is still, strictly speaking, dealing only with the expression of value: that of commodities in terms of gold; and that of gold in terms of commodities. We are dealing with instances of objectified labour weighed up against one another. But outside of the area of commodities, as bodies of objectified labour, serving as expressions of value for one another, there is the process of turning living labour into those bodies of objectified labour.

Let us assume, for the moment, that the social form of labour is identical with its natural form, i.e., labour-power itself is not a commodity. The economic question then becomes only one of the productivity of labour. This is a straightforward input-output question. Labour-power is a product (of labour) which, during its exercise, objectifies more labour than it assimilates. Productivity of labour concerns itself with the differential between the quantity of labour objectified and the quantity of labour assimilated, the latter itself consisting of objectified labour. The moment this labour-power, during its exercise, has fully re-objectified the labour-power which it has assimilated, we say that the labour-power has replaced the cost of its own reproduction. A certain portion of the labour-time, therefore, goes towards its own reproduction, this portion being directly expressed in a definite number of products. This constitutes necessary labour-time. Where labour is performed in isolation, as, e.g., where the (theoretical) labourer produces his own necessaries of life, it is immediately obvious when enough has been produced for his labour-power to be reproduced, and when that amount has been exceeded, or, indeed, when there is a shortfall. The distinction between necessary labour-time and surplus labour-time is, therefore, similarly obvious. The same applies in those forms of economy where labour is immediately social.

To return now to the world of generalised commodity production. Here labour takes the form of wage-labour and is, therefore, as much a commodity as anything else, including gold, is a commodity. The value of labour-power, like the value of everything else, is the amount of labour-time recognised by society as necessary, on the average, for its production. The expression of this value alongside other values in terms of the use-value of a single commodity gives to the value the form of price. Price is the rendering commensurable of all values in their infinite range of concrete embodiments through their being expressed as so many different quantities of a single concrete embodiment, a single commodity.

If the values of all commodities are expressed in terms of gold (price: x gold), then so must be the value of labour-power. Wage-labour is labour for *money*. That it must, during its exercise, assume this, that or the other concrete form is of only secondary importance. Through its exercise, labour-power makes the same metamorphosis, C - M, as do all other commodities. Where money is gold, wage-labour is labour for gold. The value of labour-power *in general*, living labour, rather than the value of commodities, objectified labour, now confronts the value of gold.

A rise in the value of gold means a decline in the *price* of labour-power; while, similarly, a decline in the value of gold implies a corresponding appreciation in the price of labour-power. But this rise or fall in price demonstrated by labour-power, occurs across the board, affecting all commodities with values expressed in gold, i.e., in the same price-jurisdiction, so to speak. The value of labour-power, therefore, neither appreciates nor depreciates *vis-à-vis* commodities in general. The ratios in which it commands its own means of consumption remains the same. Its value remains the same despite an alteration in its price.

> If there is a fall or rise in the *value* of gold..., in which the exchange value of commodities is measured as price, then prices rise or fall because a change has taken place in their standard of value; and an increased or diminished amount of gold... is in circulation as coin because the prices have risen or fallen ibid., p. 160, emph. orig.).

We notice that a rise or decline in the value of gold makes no difference to the value-relation between commodities which express their values in terms of that gold, i.e., in *circulation*, and that the commodity labour-power is no exception. But how is in *production*? In order that this may be better explored, we need to separate gold as particular wealth from gold as abstract wealth. This is necessary because the production of use-value I is subject to completely different laws from those governing the production of use-value II. We therefore first consider a rise in the value of gold as particular use-value, i.e., as a lump of metal, and then, separately, as universal use-value, i.e., as the material of money.

When something increases in value, it means that a greater amount of labour-time has henceforth to be spent on its production than before. It is no different with gold. A rise or fall in the value of gold means that, respectively, a greater or lesser amount of labour-time is now needed for the production of the same quantity of gold. For ease of argument, we shall confine ourselves to a rise in the value of gold, since the case of a fall requires a simple inversion. In this case more labour-time will be required for the production of the same quantity of gold. Put differently, for the same expenditure of labour-time, less gold is produced. Less gold will have been produced within that portion of the labour-time which constitutes the socially necessary labour-time for the reproduction of gold-producing labour-power than had been the case before.

In *quantity*, therefore, the gold will not attain to that formerly necessary to reproduce the labour-power. In *value*, though, this lesser quantity does so adequately, since it has been produced with the same amount of labour-time as the larger quantity before it. This is reflected in the higher value of gold in production. If we now cast gold in the role of *general* equivalent, i.e., still ordinary commodity, we find that while gold-producing labour now produces less gold for the same amount of labour-time, it is nevertheless able to continue reproducing itself as if no change has occurred, since the lesser gold commands the same basket of commodities as did the greater amount before. The important point to establish here is that a change in the value of gold *as ordinary commodity* does not affect the value of gold-producing labour-power.

That it is in the nature of gold as money-commodity that the value of the labour-power producing it be reduced to its absolute minimum has already been established. What we are about to describe applies over and above the constraint imposed by abstract wealth. Because the value of gold-producing labour-power is already at its barest minimum any increase in the value of gold must have the following results:

The increase in the value of gold, i.e., the increase in the labour-time socially necessary for its production, means that in a given amount of time less gold is produced. The portion of labour-time in which the value of labour-power is reproduced is thus increased and surplus labour-time reduced. A larger portion of the total product would have to be set aside for the reproduction of labour-power. Production will proceed upon the smallest of surpluses for the simple reason that its product is abstract wealth. But when the surplus portion has been entirely annexed by the necessary portion, i.e., the entire product is given over to the reproduction of labour-power, then production is abandoned. At such a point labour-power will have ceased to be labour-power, for the inherent ability of labour-power to produce more than it consumes will have been annulled.

This is so regardless of the social form that labour assumes. The state-run, serf-based goldfields in Siberia were abandoned in just the same way as white free diggers in the United States abandoned their claims and allowed them to be taken over by Chinese and Indian free diggers who could work them at a lower value of labour-power than even the crude white diggers of 'western' film notoriety. Del Mar's observation about California and Russia makes the point:

> The Yenesei gold regions [Siberia] were first worked by the government, employing convict labour, including women and even children. This practically meant cruelty, revolts, desertions, the knout, and the stealing and embezzlement of gold. ...In 1853, due to the falling off in returns, it ...abandoned the mines to 'private enterprise' [a system of private ownership of the enterprise based on production by unfree labour—FS], a system, which, combined with police lettres de cachet and the hiring of convict labour, increased for a time the productiveness of the mines. Under this system ...the cost to Russia ...may be briefly summed up as follows: Twenty thousand Russian lives, consigned to exile, neglect and oblivion: plus the trifling cost of keeping these unhappy creatures alive until they were worn out. The average value of the gravel was about 12 cents (6d. sterling) per ton. There are hundreds of square miles of far richer ground in California, which has never been worked and which can be had to-day for the asking, but from which American labourers turn with disdain. This computation may serve to measure the difference between a free man's ration and the meagre one upon which the Russian convicts were condemned to die (Del Mar, A., 1969, p. 385).

Or, to cite the case of free miners in British Columbia, Lock says of their claims, 'from their inaccessible position, limited character, poor pay, or depth of cover, they [the claims] have been abandoned or allowed to fall into the hands of Chinamen' (Lock, op. cit., p. 47).[5] What immediately strikes the reader from a list of all the goldmines operating in the Cariboo District of British Columbia in the late 1870s provided by Lock, is that a watershed in goldmining had been reached: though there was still gold, no further gold could be extracted unless either there was a drop in the value of labour-power, or the technology was revolutionised. Of the 45 sites mentioned, most of them had either 'not yet been proved', i.e., awaiting adequate technological application; or 'worked out', i.e., too much labour-value had to be put in for too little gold-value derived. In both cases the mines could be rendered productive by the application of new technology, but in the meantime, in both cases the mines were being worked by workers whose labour-power was of much lower value, viz., Chinese and Indians. Lock eventually does make this connection when, later, he observes about the Fraser River, British Columbia, that, 'much gold is obtained by Chinamen and Indians on the Fraser, and it is probable that eventually, many of even the higher flats and benches will pay for hydraulic work' (ibid., p. 62. The list appears on pp. 56-9). Of Illinois Canyon, California, Lock says: 'this ground has been worked by the hydraulic method, and is said to have paid a little less than wages, on the average' (ibid., p. 142).

Of course, this would all be avoided if the productivity of labour were sufficiently increased, which, from time to time, it is. But the productivity of labour in mining, an extractive industry, develops differently to the way it does in other industries.[6] Gold in its character as ordinary commodity needs to be briefly recalled. Extractive industry suffers the peculiar handicap that for every unit extracted, one unit less remains. Over time, therefore, running contrary to the general development in the productivity of labour, the productivity of labour in mining necessarily declines as the material extracted becomes more thinly spread. The actual movement in mining productivity is the resultant of these two contradictory movements.

But, in addition, the positive movement, the movement consistent with the general upward movement of productivity as characterises industry in general, is itself only partially and tardily manifested in mining. This is on account of the general inaccessibility of the resources (e.g., inhospitable terrain, deep underground, etc.) and the generally constricted space in which production takes place, as discussed above.

These objective circumstances serve to isolate the productivity of labour in mining from the average for production in general. Whereas labour in production in general can reduce the value of commodities by improving its productivity, i.e., by reducing necessary labour-time, thus both reducing the value of its labour-power *and* maintaining its standard of living, labour in mining is not so adaptable. The result is that there has been very little room for variation in the value of gold. Hemmed in between the value of labour-power on the one side and the productivity of labour on the other, there were distinct limits to the extent to which gold production could accommodate a rise in the value of gold. A *decline* in the value of gold simply means that the *quantity* of the product accruing to the labourer, or which the free miner puts towards his own consumption, will be reduced to reflect the maintained value of labour-power, rather than the new value of gold.

We return to the main question. What are the implications of a change in the value of gold-producing labour-power itself for the inter-relationships between $v_{LP(G)}$, v_G, v_C and $v_{LP(C)}$? $v_{LP(G)}$ may change for a number of reasons, mainly relating to changes in the values of those commodities, v_C, which go towards the necessaries of life of the gold producers, and changes in the standard of living of the gold producers themselves. Returning to circulation, let us assume a rise in v_C and that this rise is unrelated to any movement in the value of gold, v_G.[7] Each commodity would then command a higher price in gold than it had previously done. The total of prices of those commodities going towards the gold producer's means of living, too, would be higher. A greater portion of the product of the gold producer would therefore have to be set aside for acquiring his means of living. Necessary labour-time in gold production therefore increases to become a larger portion of the total production time than before.

What happens when $v_{LP(G)}$ rises? A rise in the value of labour-power (for whatever reason) means that a larger portion of the total labour-time will be given up to necessary labour, resulting in a diminished portion remaining as surplus labour. Once the entire former surplus portion has become part of the necessary portion, production will be abandoned (the surplus portion can, of course, be expanded by an increase in the productivity of labour).[8] A decline in the value of gold-producing labour-power means that more room is created within which the value of gold may vary, given the uncertain way in which the value of gold will in the future be affected by the richness/poverty of the source. The early South African goldmining industry illustrates this point well, as is examined in Part III.

A distinction has to be made between gold in its aspect as measure of value and gold in its aspect as standard of price. This is necessary because these place different demands on the *physical* form of the material. In its aspect as measure of value, gold needs present itself only as a homogeneous, stable material infinitely capable of division and re-unification. It is unimportant whether this gold appears physically as dust, nugget, jewellery, coin, bullion, or anything else. Indeed, the more amorphous its physical form, the more adequate is it to this role. The lack of specificity of its physical form most adequately expresses the lack of specificity of the labour it has to measure. It must, of course, *socially* itself be a thing of value.

Standard of price, on the other hand, has its own, quite different, set of demands. Since the value of commodities is measured in the physical body of a single commodity, a definite quantity of that commodity must stand as the unit of price. And, like all units, it immediately posits a system structured around itself as central calibration, with fractions below and multiples above it. In the case of gold this unit is a unit of weight. The expression of the values of commodities are now measured in terms of unit quantities of gold. The value of gold assumes a mere relative form while its use-value serves as equivalent. 'Gold as materialised labour-time is a measure of value, as a piece of metal of definite weight it is the standard of price' (Marx, 1977b, p. 71).

The standard of price is essentially 'internally directed' (for want of a better phrase) since unit of weight serves for the systematic expression of different *quantities of gold* as fractions or multiples of a central quantity which is a clearly identifiable and easily comprehensible unit. The infinite range of *commodity* values, expressed in an infinite range of quantities of gold, can thus be meaningfully related to one another as representatives of the values of commodities, rather than the value of gold.

A unit of weight remains a unit of weight, and any fluctuation in the value of gold will merely give a different value to the unit of weight. The latter, therefore, need not change in response to a change in the former. The standard of price is therefore stable, acting as a kind of damper between the value of the measure of value and the values of the various commodities needing to be measured by it. 'Gold is the measure of value because its value is variable; it is the standard of price because it has been established as an invariable unit of weight' (ibid.).

This is the contradiction of gold: ideal fluidity and ideal fixity coexisting in the same physical thing. This begins to show itself the moment we resituate standard of price into circulation (although it must be remembered that standard of price functions independently of whether the material of that standard actually circulates). The unit weight of metal which is the standard of price is rendered adequate to circulation by the metal being divided into discreet bodies, some having that unit weight, others fractions of that unit weight, and still others multiples of it. Initially, i.e., historically, they had their respective weights marked on them for ease of recognition. They also came to be shaped into discs for ease of handling. Thus does the physical shape of the material gold, at this point, administer to the needs of circulation by adopting the form of coin, and of coin of different denominations.

When we say that gold is the money-commodity, it means that all other commodities express their exchange values as particular quantities of gold. The values of various commodities are here compared to the value of gold, that is true. But for so elementary a purpose there is no need for a universal equivalent. The elementary expression of value would suffice. The purpose of expressing value in an universal equivalent form has, in fact, nothing to do with the value of the particular commodity which serves as universal equivalent. The universal equivalent is posited by *commodities in their exchange with one another*. That the exchange value of commodity A appears as x oz. of gold, serves only to compare it with the y oz. of gold which expresses the exchange value of commodity B. Since it appears as if commodities are comparing their values with the value of the universal equivalent, whereas they are, in fact, comparing their value with one another, and since one ounce weight of gold will always be one ounce weight of gold, the standard of price parts company with the measure of value. It, therefore, does not matter what the value of gold is, its variation will merely mean that, all else remaining the same, commodities will exchange *for one another* in greater or lesser quantities at the given unit weight of gold but the ratios in which they exchange will remain the same. In its capacity as standard of price, the demand on gold is merely that it has a value, not that that value be of this or that magnitude. As measure of value, it has to have a definite magnitude of value.

Since commodities exchange for one another and not for gold, and the actual value of gold is irrelevant in its role as standard of price, the value of the unit of gold which serves as standard of price, does not have to reflect the actual value of that unit of gold as measure of value. The function of standard of price would be equally well served by coin, paper, strips of leather, etc., provided only that gold remains the measure of all values. Gold can, therefore, be withdrawn from circulation and replaced by paper tokens with national names printed on them.[9] The paper currency (or other token), when first introduced, would be a simple replacement for the unit of gold serving as the standard of price, in this case the ounce: x units of paper currency = 1 ounce of gold. The amount of value represented by such a token *as standard of price* is the value of commodities expressed by the unit of gold which that token represents. It thus becomes a surrogate expression of the prices of commodities and as such, of the values of commodities.

> A definite quantity of gold as such does not express a value relation, nor does the token which takes its place. The gold token represents value in so far as a definite quantity of gold, because it is materialised labour-time, possesses a definite value. But the amount of value which the token represents depends in each case upon the value of the quantity of gold represented by it. As far as commodities are concerned, the token of value represents the *reality of their price* and constitutes a token of their price and a token of their value only because their value is expressed in their price (ibid., p. 115, emph. orig.).

If an actual ounce of gold were the reality of the price of the commodity, then a token of that gold is a mere token of that price. But since it is the price of *the commodity* and not the price of gold, this token now appears to distance itself from gold, where it appears not to belong, and conceptually appears to move closer to the commodity, where it does appear to belong.[10] This token of value, not being value itself, has no existence other than as a means of purchase, as a representative of the universal equivalent in this narrow specificity. It does not itself circulate as value. The actual value of this token, the labour-time necessary to produce it, is therefore as irrelevant as that of gold as standard of price. Indeed, the token of value becomes a more adequate expression of the standard of price. the less value it itself has — hence paper currency. Its quantity in circulation does not, therefore, depend on the total of prices needing to be circulated, but can be any arbitrary amount.

This is unlike where gold directly circulates, i.e., where it functions directly both as standard of price and as measure of value, even though it is present in circulation only in the former and not in the latter. The rate at which paper notes will represent value will depend on their quantity in circulation. To put the matter differently, since these ideally valueless tokens do in circulation represent gold, their *quantity* will depend on the quantity of gold that would otherwise circulate, and since they are tokens of *value*, the amount of value they are able to command depends on their quantity. 'Whereas, therefore, the quantity of gold in circulation depends on the prices of commodities, the value of the paper in circulation... depends solely on its own quantity' (ibid., p. 119). In other words, each unit of paper currency becomes a fraction of a definite total quantity of value needing to be circulated. The total value represented by paper currency must, for now, represent this total quantity of value.

Whatever the value it represents may be, though, it must be constant long enough for it to become generalised throughout the network of exchange relations and to be a reliable bearer of value from sale to purchase and vice versa. The custodian of the form of economy, the state, which issues these tokens in the first place, also ascribes to them a legal rate at which they will represent gold. This legal rate at which it *represents* gold is, initially, also the rate at which it is *convertible into* gold. Thus has the reality of price as a converted form of value at first to be acknowledged if a token is to be the instrument of that conversion.[11] Although physical gold will, therefore, continue to circulate as coin, there is, in essence, no longer any need for it to do so. The token will gradually replace it as medium of circulation not only because the former is a more efficient medium than the latter, but because physical gold can then freely withdraw into that aspect of money to which its physical form is most adequate: value-for-itself, or store of value.

We must now make a lengthy digression to introduce paper currency and the contradiction between capital and money as it emerges in the relationship between gold and paper currency. This is necessary in order to explain the imposition of a 'fixed price' on gold together with fixed rates of exchange between currencies. This arrangement, it will be shown, contradicts the very nature of gold as money-commodity.

The 'fixed gold price' vs the value of labour-power

While circulation unquestionably develops forward with the introduction of paper currency, such a move, at the same time, represents a retrogressive step. The hitherto-existing direct circulation of gold in the domestic economy, meant that all domestic economies had the same currency. Their respective average levels of productivity were directly able to influence one another through the conduct of international trade.[12] The same gold which bought a kettle in a village in North Yorkshire could buy a sackful of grain in Italy.[13] Paper currencies are strictly national money, and their circulation isolate domestic economies from one another.

Paper currency, convertibility and world gold movement

Changes in the average level of domestic productivity were reflected in changes in the general level of prices *in the local currency*. And since the currency was only a standard of price and not a measure of value, there was no way of comparing international productivity levels, and no way of knowing whether a particular commodity was more cheaply produced in one country as opposed to another. The matter is further complicated by the fact that each state set the rate at which its paper currency would represent the unit weight of gold only with reference to itself. If all national currencies were set in such a way that a single unit of each currency represented the same unit weight of gold, then any changes in national price levels would still be discernible internationally. But this was not so. Each national currency had its own datum, so to speak.

International trade now required a mechanism for providing buyers in one country with the means of purchase in another. The foreign exchanges, though, were not exchanging values, but tokens of value. Their rate of exchange internationally depended purely on the extent to which they were in demand. By their nature, therefore, paper currencies lend themselves to wholesale speculation, and the essential quality of the medium of circulation, that the value it represents endures throughout the act of exchange, disappears.

If this is a problem for the circulation of commodities, it is even more so for money. In the same way as the circuit C - M - C obscures the object of commodity circulation, C - C, so the circuit M - C - M obscures the object of money circulation, M - M, indeed, M - M'. Preserving value for the duration of the act of exchange, i.e., from sale to purchase, is a problem which commodity circulation solves by settling upon gold as its medium. We have seen how the conscious hand of the bourgeoisie, the state, has to contrive a device for ensuring such preservation once the further development of exchange itself marginalises gold from this role. *Enhancing value* (as opposed to merely preserving it)[14] during the course of the act of exchange, i.e., from purchase to sale, is a problem which money circulation solves by money positing capital. Marx explains it thus:

63

> Posited as a side of the relation, exchange value, which stands opposite use-value itself, confronts it as money, but the money which confronts it in this way is no longer money in its character as such, but money as capital, (Marx, 1973, p. 269).

The circulation of tokens of value, which we shall call **T**, in the national economy compels the circuit **M - C - M'** to assume the form of **M - T - C - T' - M'**, where two metamorphoses have become four. Two of these, **M - T** and **C - T'**, involve, in fact, a complete dissolution of not only the form of value, but of value itself, while the other two, **T - C** and **T' - M'** amount to the restoration not of particular forms of value, but again of value itself. Already, in the 'mere' act of introducing paper currency, not only is a more efficient allocation of social labour effected, but value begins to abolish itself, at such points to be replaced by conscious decision-making at the level of the society as a whole. But this development, being a moment of the contradiction inherent in the commodity, that of use-value and exchange value, has here, too, its opposite pole. For money, **M**, this is a gauntlet, and it must seek to guarantee its token, **T**, as an adequate representative of itself. Thus must conscious social action turn on its head and immediately salvage that which it has just annulled. Thus does the state, that most visible hand of the bourgeoisie, institute a fixed rate at which **T** will always represent **M**.

No sooner is such a rate fixed, and the metamorphosis **T' - M'** can be dispensed with, or at least, suspended. **T' - M' - M - T** becomes superfluous, since **T'** is by reflux already **T**. To effect the transformation **T' - M'** is to depart from circulation and the moment money departs from circulation, it ceases to function as capital. The object now becomes to avoid **M** altogether so that the circuit **T' - C - T'** may be repeated over and over again.

> Since merchant's capital is penned in the sphere of circulation, and since its function consists exclusively of promoting the exchange of commodities, it requires no other conditions for its existence—aside from the undeveloped forms arising from direct barter—outside those necessary for the simple circulation of commodities and money (Marx, 1984, p. 325).

There thus appears to be no further need to maintain the convertibility of **T'** to **M'**, and hence of **T** to **M**. The concept of inconvertibility is inherent in the concept of convertibility. As money, value embraces convertibility; as capital, it shuns it. The very protection which value seeks for itself as money becomes its barrier as capital.[15] 'It is evident that banks issuing notes can by no means increase the number of circulating notes at will, as long as these notes are at all time exchangeable for money' (Marx, ibid., p. 523). In other words, convertibility is a constraint on capital (but in more ways than one). This is a contradiction between two forms of value: money and capital.

The fragmentation of circulation into distinct national components each with its own medium of circulation, heightens this contradiction. Let us return to the domestic circuit $M - T - C - T' - M'$. The same basic processes which occur in the domestic circuit occur internationally, except in a more complex manner. An import of commodities is, at the same time, an export of money; though an export of money is not necessarily an import of commodities. We shall first examine the import and export of commodities, then the import and export of money, which may be either as means of purchase, means of payment or capital.

The above circuit, as it is extended across more than one country, may be given as: (i) money export-commodity import:

$$M - T_1 - T_2 - C - T_1' - M'$$

(ii) commodity export-money import:

$$M - T_1 - C - T_2(T_1') - T_1' - M'$$

With respect to (i): Money (gold), M, is converted to local currency, T_1, at the central bank; local currency, T_1, is converted to foreign currency, T_2, at the foreign exchange (either locally or in the foreign country); foreign currency is exchanged for commodities, C, in the foreign country; the commodities are imported and sold for local currency, T_1; which is, in turn, converted back to gold, M, at the central bank. If M has proved successful as capital, then the imported commodities would sell at local currency T_1', which would become M'.

In the case of (ii): Money (gold), M, is converted to local currency, T_1, at the central bank; with this local currency commodities, C, are bought; these commodities are exported and in the foreign country sold for foreign currency, T_2; this foreign currency is converted back into local currency, T_1, (either in the foreign country or locally); and the local currency is redeemed in gold, M, at the central bank. Again, if M has proved successful as capital, then the foreign currency, T_2, would be the equivalent of an enhanced local currency, T_1', which would become M'.

All of these metamorphoses have already been discussed in respect of the domestic circulation of convertible paper currency, above, except for the conversion between two national currencies, $T_1 - T_2$ and $T_2 - T_1$. It has been argued above that the foreign exchanges do not exchange value, but tokens of value. While it is the state which fixes the legal ratio in which paper currency represents a given weight of gold, it is also the state which creates paper currency. Because their quantity in circulation is not determined by the total of prices needing to be circulated, any quantity may be printed. The rate at which each unit will exchange against commodities depends on what fraction of the total prices is accounted for by that unit.

The money-dealer inserts himself between the demander of foreign currency (the latter being also a supplier of local currency or gold) and the supplier of foreign currency (who is at the same time a demander of local currency or gold). The money-dealer's own capital follows the circuit:

$$M - T_1 - T_2(T_1') - T_3(T_1'') - T_4(T_1''') \ldots - M'$$

Again, as soon as money is represented by any token, it can quite happily remain in that condition, increasing its power of *representation*, rather than of realisation, with every transaction. Once money leaves the physical body of gold behind it, it will delay its return to that form in order to remain in circulation, where it continuously increases its *claim* on gold, without ever realising that claim. This holds true also where the token exchanges for commodities, rather than other tokens. Convertibility assists money in its becoming capital; but once capital, it, in turn, drives money to inconvertibility.

But the rate at which T_1 will exchange for T_2 is not only a simple function of supply and demand. For the same two currencies there would be different rates of exchange depending on whether the transaction is conducted in the one country rather than the other; and also depending on whether the transaction takes place today or tomorrow. Uncertainty as to the 'real' rate of exchange — as it expresses different productivities of labour and different intensities of trading activities in the different countries — is therefore compounded by an exponential factor.

This has the effect that not only would money go into circulation for shorter periods, but that it would also restrain itself from venturing too far along its circuit of formal transformations. There is a reluctance on the part of the owner of capital to tie up his capital for too long in forms which cannot be easily reconverted into gold. The many uncertainties brought about by the existence of many national currencies is therefore a direct impediment both to money and to capital.

In the exchange of commodities between national economies, some countries, after a definite period, find themselves in possession of commodities which they have not paid for, while others find themselves divested of commodities without having received payment for them. This in itself amounts to a sophisticated clearance system, where actual circulation of the money material is confined to the settling of balances. Unfortunately for the policy makers, the movement of gold into or out of a domestic circulation area interferes with the functioning of the convertible national currency. Its volume has to be adjusted so that it maintains the legally required ratio to the national gold reserve. Currency was injected into or withdrawn from circulation by central bank interest rate manipulation. Higher interest rates reduced the level of currency, lower interest rates increased it.

But this very device turns the currency into an object of capital. Monies which might in a former age have petrified into hoards, find here one more opportunity for expansion, a chance to realise the essence of money — to become capital. A third international movement of value, in addition to those of goods and money, is now a movement of capital. This created its own demand for national currencies quite separate to the demands of trade or those of the money dealer, with its own influence on currency exchange rates.[16] Such capital movements, too, occurred in the form of gold shipments. The result is that the very material by means of which national governments sought to stabilise the relationship between their domestic economies and the world economy, was the very instrument of instability. As money, gold enabled capital, as capital gold disabled money. By international agreement, the commercial powers of the world sought to stabilise money by instituting a system of fixed exchange rates based on a fixed rate at which certain key national currencies may be redeemed against a definite unit weight of gold. In other words, a single international standard of price was put in place. This was called the international gold standard.

Our purpose is solely to expose the implications of the International Gold Standard system for the value of gold-producing labour-power. This is done firstly by simply describing empirically the workings of that system. We then see whether its workings can throw any further light onto the nature of the money-commodity. This chapter will show that the International Gold Standard system merely gave institutional expression to the contradictions inherent in the commodity, and which are developed to a higher degree in the money-commodity. The contradiction between use-value and exchange value, which is *solved* for the money-commodity, gold, is reinstated for it by the gold standard on account of its inability to accommodate the dual character of gold as both commodity and non-commodity. At this point, the so-called 'fixed gold price' emerges as a fixed gold *value*. The latter set the limit beyond which the value of gold-producing labour-power could not appreciate.

The International Gold Standard system[17] consisted of the following institutional arrangements:

1. The national currencies of member countries are directly convertible to gold at a fixed currency-to-gold ratio[18] and in unlimited amounts; 2. the ratio shall be practically the same regardless of whether the state acts as seller or purchaser of gold; 3. all member states shall permit unrestricted import and export of gold (Day, A., 1960, p. 438, or, indeed, any textbook which mentions the gold standard).

Although it was immaterial to the principle of the system whether gold actually circulated in the domestic economy (gold specie standard), or whether this role was played by a fully-convertible paper currency (gold bullion standard), this did make a big difference in the practical operation of the system, since, the arrangements listed above require each member country to hold, in fact, *two* gold reserves: one for domestic; and the other for international purposes, if it did not want its own circulating medium controlled by the international movements of gold.

Where gold directly circulates in the domestic economy, the domestic reserve would, strictly speaking, be the same body of gold as the gold in circulation. The volume of gold in circulation would contract and expand in response to the need for gold in the economy. This would be the sum of, on the one hand, a function of the total of prices to be circulated and the velocity of circulation, plus, on the other hand, the total amount of gold issued as money capital. Where gold does not circulate, but is represented by a fully-convertible paper token, the domestic reserve should vouch for the entire sum of paper currency. This paper currency would then, naturally, require its own *paper currency reserve*, for as it expands and contracts, it will be expelling from circulation such amounts of currency as become, from time to time, superfluous, and recalling back into circulation such varying amounts as and when needed. Since the currency remains fully convertible, the domestic gold reserve may be described as semi-active. It nevertheless has to be kept.

As far as the international exchange of commodities were concerned, payment was made either in the currency of the country selling, or in gold.[19] Here we might describe a country's international gold reserve as a reserve for money. Realisation in commodity circulation, whether domestic or international, consists in the owner of value in the form of commodity A eventually being the owner of that same value in the form of commodity B. Realisation in money circulation consists in a certain amount of money, M, being converted into a higher amount of money, M'.

The same problem of the medium of circulation enduring *as a value* through the act of exchange arises here too, only more starkly so. The duration of the circuits C - M - C and M - C - M are not only longer, but also more complex, as discussed above. Their longer duration allows greater opportunity for the rate at which different national currencies exchange for one another to change, thereby increasing the chances of non-realisation. Hence the need to fix the rate at which currencies convert into one another and into gold. But this conversion of currency into gold is also a conversion of gold into currency. This reversibility of the relation is purely formal, though, since paper currency remains, essentially, a token of the value of gold. Value, a thing which is by its nature variable, is now subjugated to the standard of price, a thing which is by its nature fixed.

We now return to the value of labour-power in gold production. The latter half of the nineteenth century saw a dramatic rise in the average level of productivity in the industrial economies and hence a decline in the average value of labour-power. Particular contributors to this were, e.g., the decline in transport costs afforded by railways and steamships, the substitution of high-quality steel for iron and a grain market which spanned the globe. Not only fixed costs, but variable costs, too, declined.[20] Although commodities exchanged for tokens of value and not for gold directly, these tokens were nevertheless convertible to gold at a fixed rate. In general, the values of commodities would not change in relation to each other (of course the rate of productivity increase happens differently for different commodities, but their values were all moving in the same direction), but they would change in relation to gold. Assuming the value of gold to have remained constant (this was not the case), then commodities would become cheaper in terms of gold, while gold became more expensive in terms of commodities.

But the *production* of gold was not subject to the same movements in productivity as was its *circulation*. The first reason for this has been discussed above, and relates to the isolation of the productivity of labour in extractive industries from the general movements of productivity in industry in general. In extractive industry productivity gains following on the introduction of new technology tends to be offset by increasing difficulty in maintaining the volume of production of the material being extracted.

This isolation receives further impetus from the actual geographical isolation of gold production from the centres of industry and the technological backwardness of gold-producing areas when compared to industrial areas. The average level of values is higher in mining areas in general, than it is in the industrial centres. The underlying tendency is therefore that the values of the products of extractive industry will, on average, be higher than those of industry in general.

To the extent that capital is restrained from flowing freely in and out of industries, and the products of particular capitals from having to compete for sale against one another, a rent will emerge. Prior to capitalist mining (viz., the ground owned by the owner of the capital) the surplus-value eventually realised through the sale of the product had to incorporate a portion to be handed over to the landlord. With the rise of capitalist mining in the late nineteenth century, this rent disappeared.[21] Monopoly rent, i.e., the difference between a monopoly price and the ordinary market price, persists in mining in general. Such monopoly rent may arise on account of: (i) a more favourable location or endowment with richer ore — either way, amounting to an increased productivity of labour which has nothing to do with the organic composition of capital; or (ii) 'the purchaser's eagerness to buy and ability to pay, independent of the price determined by the general price of production, as well as by the value of the products' (Marx, 1984, p. 775).

Goldmining has its own, more elementary response to differences in productivity between different enterprises, and between itself as an industry and industry in general. The 'fixed price' of gold means that sudden increases in the productivity of labour do not translate into monopoly rent, instead, the 'latitude' provided by this increased productivity is taken advantage of to extract a lower grade of ore than would normally be economical to extract. What is potential value one day becomes actual value the next. The fixed price serves as a barrier to the competitive transfer of surplus-value between gold-production and other industries. Such transfer can only take place indirectly, as social capital. This is discussed in more detail under *Competition in gold production*, below.

Changes in the value of gold does not affect the rate at which commodities exchange for one another. Hence neither does fixing the rate of convertibility of paper currency into gold at a certain point affect the rate at which commodities exchange for one another. Fluctuations in the values of commodities relative to one another are reflected in their changing relative prices. Gold here officiates in its capacity as standard of price although it does not circulate. Its value is represented for it by paper currency.

Where gold does, however, directly circulate, it must represent its own value. This it does (i) between the boundaries of the circulation areas of the various national currencies, and (ii) at its point of entry into circulation, the point at which it leaves production. In neither of these areas, however, was it able to do so. Although gold circulated physically between countries, it now no longer circulated as value. The material gold now performed a function *for* paper currency: It went where paper currency could not go. *Real gold now became a representative of paper currency.* Thus does the token of value, posited by gold, now posit gold as a token of it. (ii) At its point of production, gold enters into its social intercourse not as *money*, but as *commodity*. And commodities are circulated by a token of value, paper currency. When Marx speaks of the contradiction between use-value and exchange value having been solved for gold (Marx, 1977b, p. 48), that solution is annulled when gold, as it emerges from production as commodity, does *not* barter, as Marx contends, but is *assigned* a price, one quite unrelated to its value. The token of value can represent only *a fixed quantum of value*, which, after all, is why it can be a convertible token. Although the values of commodities find their fluctuating values expressed in fluctuating paper currency prices, it is not so with gold. The rate at which gold exchanges for paper currency is fixed in the capacity of gold as *money*. That *very same material* now confronts paper currency as *commodity*.

> In so far as gold is to be established as the unit of measurement, the relation of gold to commodities is determined by barter, direct, unmediated exchange... With barter, however, the product is exchange value only in itself; it is its first

phenomenal form; but *the product is not yet posited as exchange value* (Marx, 1973, p. 204, emph. ours.).

In other words, gold is here commodity, but it is not yet money. But the token of value, paper currency, makes no such distinction. It knows gold only as a material. It recognises only its weight, and not its social form, although paper currency is itself a product of that very social form. When it represents the value of gold, that gold is something which *effects* exchange, not something which *participates* in it.

Since nothing can be its own equivalent, the token of value cannot do for the commodity gold what it does for other commodities. It cannot give a price form to the value of gold, as it does to all other commodities. While the standard of price is embodied in something other than the measure of value, it must have its relationship to that measure defined by the state. This relationship is one of representation, and representation at a definite ratio. While, for whatever reason, the state has to maintain this ratio unaltered over time, the token of value has no authority to represent any more than a fixed maximum of the value of the material which serves as measure. To the extent that the value of the measure of value, i.e., gold, exceeds this maximum, gold has no mechanism through which to assert its value equivalence against other commodities. Fluctuations in the value of gold can thus not be reflected in the token.

The producer of gold must, therefore, produce gold *below* this value or not at all. This holds equally for whatever social form that production might assume. It makes no difference whether that gold is produced by slaves, free diggers, wage-labourers or a co-operative, whether the enterprise is run on the basis of one man or as a vast-scale venture, whether production is controlled by the state, private capital or joint-stock capital, and, regardless of whether production takes place in the same country as that where its product is to serve as money, or in a different country.

This does not mean that the value of gold cannot fluctuate. It is produced by labour, therefore it has a variable value. Where production is carried on under value conditions, variable value (that expended on labour-power) and fixed value (that expended on equipment, raw materials, etc.) fuse to confront the producer generically as his cost of production. All producers seek to reduce their cost of production, although this assumes a peculiar form in capitalist gold production, which is discussed under competition in goldmining, below.

Thus there arose, in the final quarter of the nineteenth century, a situation in which the organic link between the value of the money-commodity and the values of commodities in general was severed. The value of gold as it entered into circulation became, effectively, fixed by the fixing of the ratio in which any one national currency was allowed to represent a unit weight of gold, and the fixing of the ratios in which various national currencies were permitted to take the place of one another as representatives of that gold.

Since the richness with which gold occurs in nature remains highly subject to chance, its value, independently on this account, must remain variable. The room required for this variability can only be provided by not allowing the value of gold-producing labour-power to appreciate. That which is by nature *variable*, must now become *fixed*. How these developments played themselves out empirically is explored in Parts II and III.

Notes

1 For a discussion of the relationship between necessary labour-time and surplus labour-time, both in general and in its particular expression under capital, see Marx, 1973, pp. 704-712.
2 It is this which underlies Jacob's doubt 'whether gold has ever been paid for at its full value', and not the amount of value that goes into its production, as Marx seems to think. His linking of diamonds with this speculation of Jacob is inappropriate (see Marx, 1983, p. 47). The way in which to understand Jacob's remark is that the gold-producer has a tendency not to return to labour-power (whether his own or that of others—prior to wage-labour, that is) the *value* it has expended on gold-production, but less than that value. Siculus' observation of silver miners being worked to death is a particular expression of this.
3 This might, perhaps, be the first form of parasitic production.
4 See the discussion of the precious metals in Marx, 1977b.
5 'The cañon near the 49th parallel,...soon abandoned by whites, was worked for years by Chinamen', ibid., p. 52.
6 Marx makes this same point about extractive industry. However, he includes also agriculture (in the broad sense) as

extractive industry. This is wrong for two reasons. Firstly, although it is correct to say that the productivity of labour in agriculture is subject to the vagaries of nature, e.g., poor soil, crop failures, etc., the product of the soil is not a finite resource. It is regenerated in that every year new seeds are sown and new crops harvested. Even the soil itself is not a finite resource, as barren or exhausted soil is brought or restored to vitality by particular methods of agriculture and soil treatment techniques. The second reason, and perhaps the more important, is that all of industry has its basis in mining and agriculture. If they were both extractive, i.e., operating on continually diminishing resources, it would be difficult to explain the evolution of production beyond mining and agriculture (see Marx, K., 1977b, pp. 37-8).

7   Again we shall confine ourselves to examining only a rise in $V_{LP(G)}$, as a simple inversion of the argument would suffice to explain a decline.

8   Such a universal increase in the value of labour-power did, in fact, take place in the middle third of the nineteenth century with the abolition of slavery and the universalisation of wage-labour. Overall, however, it is likely that the second half of the nineteenth century saw a *decline* in the value of labour-power as the general productivity of labour soared. Goldmining labour-power, though, was subject only to the rise and not the decline in value. This is taken up in Section II.

9   'In many states notes or paper-moneys are rapidly usurping the place of coins. ...The general progress of the note-system is shown by the fact that in the early part of the 19th century notes formed about one-fourth of the entire circulation of the Occidental world. At the present time [1900] notes form rather more than one-half' (Del Mar, op. cit., p. 453).

10 'But the appearance is deceptive. The token of value is directly only a *token of price*, that is a *token of gold*, and only indirectly a token of the value of the commodity' (ibid., emph. orig.).

11 Of course, the state is in a position to ignore its own rules.

12 In addition to gold, bills of exchange (a credit instrument) also served as medium of international circulation already in the Middle Ages.

13. When gold circulated as currency in domestic economies, coins from the most diverse realms circulated side by side in any one country, especially where the mint was simply too remote. What mattered was that they were all of gold. Their respective values were recognised by their respective weights. In closer proximity to the mints, foreign coins were less readily accepted and had to be melted down and reminted.
14. Indeed, its enhancement *is* its preservation, for mere preservation (as in C - C) removes the point of money circulation.
15. When the 'Great Anti-Fascist Barrier' came down in 1990, it was those protected behind it who streamed *out*.
16. For Marx's discussion of exchange rates, which was not yet a system in his day, see 1984, pp. 317-22, pp. 568-9, pp. 574-84.
17. Opinion is divided as to the date of inception of the International Gold Standard, but most writers put it to 1873 when Germany went onto the standard. Some, though, do suggest the fixing of the value ratio of the pound sterling to a definite weight unit of gold by Isaac Newton in 1717 as the birth of this system.
18. The phrase 'currency to gold ratio' is deliberately chosen. All economics textbooks, departing as most of them do from the quantity theory of money, conceptualise this ratio as a 'gold price', which makes it synonymous with the price form of the values of commodities as they are expressed in particular quantities of the use-value of the universal equivalent. It is, therefore, a designation which does not illuminate. A gold price does, indeed, emerge, but this is secular, i.e., by no means the same order of category as price, the unified expression of value.
19. We ignore the bill of exchange as medium of circulation. Although this was a credit instrument and, as such, an economiser on the circulating medium, it is not a form posited out of the evolution of money. As a form, it has no inherent relation to gold. A lengthy history of the bill of exchange is offered in Powell, E., 1966, and also Kindleberger, op. cit.
20. This 'second industrial revolution' is well documented, and is discussed in detail in Section II.
21. The depletion of the resource itself encourages this development. 'He [the landowner/capitalist mine owner]

only gives himself permission to exploit the mine. This does not enable him to draw a rent, but it does enable him to exclude others and to invest his capital in the mine, with profit', Marx, 1975, p. 362.

3 Gold as capital

Gold has been examined as use-value and as money. We must now examine gold in its specific nature as capital. By the late nineteenth century, gold has, of course, already functioned as usurer's capital, merchant's capital and money dealer's capital. The material itself only becomes an object of capital with the creation of a world market with its different productivities of labour in its different national spheres. The insertion of the material gold into the circuit of productive capital as commodity-capital is the new development of the late nineteenth century. It is this form, in particular, which we are interested in examining since it is this form which relates directly to modern gold-producing labour-power.

Chapter One has explained that once the role of money-commodity is ascribed to gold (or any other commodity, for that matter), then the same material, gold, assumes two, quite distinct use-values. One of these, particular use-value (what we called Use-Value I), is of the same category of use-value as that of all other commodities. It is a particular object resulting from the objectification of particular concrete labour. As such it addresses a particular need of a particular individual. The object of its production, ultimately, is its private consumption. Its second use-value, universal use-value (Use-Value II), does not address any particular need of any particular individual, but a universal need of exchange economy. Its consumer, therefore, is exchange itself.

All commodities, including gold, are non-use-values to their owners and use-values to their non-owners, which is why they exchange and cease to be commodities. The money-commodity, i.e., gold as Use-Value II, is a use-value to both its owners and its non-owners. Here its use-value *is* its exchange value. Its use-value II is its exchange value I. As Use-Value II, its owner consumes its utility the moment he exchanges his money for Use-Value I, i.e., as soon as he buys something with it.

77

But as Use-Value II it persists. It continues to be the material gold even after something has been bought for it. Its utility as exchange value, therefore persists. Unlike Use-Value I, which is realised in consumption, Use-Value II is realised in circulation. Where Use-Value I lacks, there is consumption restricted; where Use-Value II is deficient, there is circulation restricted (we are not yet speaking of credit).

As soon as a symbol of the medium of circulation begins to circulate alongside or in the place of that medium, and is at all times fully redeemable against that medium, everything is in place for the commoditisation of the medium of circulation *as such*. To find the genesis of this, we must look at the most elementary token of value, specie coins.

A gold coin is a fixed quantity, i.e., weight, of the metal which at the time of its production contained a certain amount of value. Coins, for ease of reckoning, appear in fractional and multiple denominations. The infinite divisibility and ability to be re-united has now been codified and restricted to these denominations. Subtle variations in the value of gold can now only be reflected in the quantity of commodities commanded by each denomination. The essential relation is now, in effect, reversed, with commodities reflecting fluctuations in the value of gold, rather than the other way round. In this situation the medium of circulation itself stands as relative value across from particular commodities, its various equivalences. Essentially the condition for trading in the medium of circulation *as if it were an ordinary commodity* is now in place.

But gold cannot trade as an ordinary commodity until another single item can stand across from all other commodities, including gold, to officiate as their equivalence. The symbol of the medium of circulation, paper currency, takes up this role. Gold can now be bought and sold by paper currency. A potentiality inherent in use-value II, that of exchange value II, now rises to actuality, and gold becomes, for a second time, a commodity.[1] While gold remains merely a commodity with an additional, social use-value, it is money. When that social use-value assumes its own exchange value (quite apart from the exchange value which it has in its character as ordinary commodity), then gold has become capital.

Given that the same item, the material gold, came to be demanded in each of its three different characters — as use-value I, as money and as capital, it follows that there would simultaneously be different degrees of demand for it. It can therefore be bought as one form and then sold as another. It is therefore only natural that, in spite of its fixed price, it would develop a daily-fluctuating premium.[2]

Here gold as capital remains of the order of merchant's capital, M - C - M'. As such it is confined to circulation and is involved only in the redistribution of surplus-value. But gold also becomes the object of productive capital, M - C . . . P . . . C' - M'. It is here that it is most relevant for this thesis, for here capital must engage gold-producing labour directly.

Gold production and the circuit of capital

In regarding the circuit M - C - M', we are normally only concerned with M which becomes M', i.e., with capital. C, the commodity, does not interest us because it enters capital's circulation process as and when summoned by capital and drops out of such circulation when dismissed by capital. C is therefore only of economic interest insofar as it circulates capital. C is normally a real use-value (either a product of labour or labour-power itself). It is created prior to its entering the circulation process and after its departure it is consumed. Capital, however, carries on to engage other commodities as it reproduces its own circulation by the proliferation of further circuits.

In the relationship between money and capital and the ousting of gold from the money function, the C in the circuit assumes a more illustrious role.[3] While in the circuit C - M - C, M was retained for the purposes of acquiring the second C, it is now, in M - C - M', pressed into service as capital. Since the objective of M is M', there would be two basic ways of achieving this: M - C - M', as above; or M - M' directly.

In the basic configuration of the circuit M - C - M', viz. as it appears here, C is a product of labour, objectified 'dead' labour. In this instance capital merely appropriates surplus-value. In one of its more complex configurations, M - C . . . P . . . C' - M', the C consists partly of objectified and partly of living labour (labour-power). In this case capital both creates and appropriates surplus-value. It is assumed that C goes into particular consumption, ΔC (added commodity) having become ΔM (added money) to be joined to M to restart the circuit.

In another of its more complex configurations, M - C . . . P . . . C' - M', C again consists of both objectified and living labour, except that the objectified labour distilled at the end of the production process coincides with that objectified labour which serves as the universal use-value, money. It therefore emerges simultaneously as both an object for particular consumption and an object for universal consumption. On the one hand, it is an object of consumption for the living individual (living labour), on the other, it is an object of consumption for capital, i.e., as commodity capital and money capital at the same time. This role as object of consumption for capital, i.e., dead labour, is grounded in its role as object of consumption for the individual, i.e., living labour.

In the configuration $M-C...P...C'-M'$, C remains value in the commodity form and as such unable to restart the circulation process anew. It still needs to transform itself from C into M. However, in the configuration $M-C...P...M'$, while value here, too, emerges in the commodity form (C' *is* M'), this commodity here already effects the transformation of C in $M-C-M'$ into M in that C emerges as M directly out of the production process. Its next action is therefore not to close the circuit $M-C-M'$ by transforming into M', but to open a circuit anew, that action being the starting point, rather than the end point of the circuit. This is the circuit of such capital as is applied to the production of a money commodity. This circuit, though not yet adequate to the ideal capital, as expressed in the circuit $M-M'$, approximates it more closely than does $M-C...P...C'-M'$.

In both $M-C-M'$ and in $M-C...P...C'-M'$ capital must twice change its form: from money into commodity and from commodity back into money. In the case of $M-C...P...M'$, it needs make only one such transformation: from money into commodity, since the second transformation is effected *within* the production process itself. Thus do we find not only the production process taken into the circulation process of capital, as in $M-C...P...C'-M'$, but the circulation process of capital taken into the production process $(...P...M')$.

This latter transformation, $...P...M'$, of course, presupposes: (i) that C which enters the production process consists of means of production (M P) and labour-power (L P); and (ii) that enough of M P and L P are added to the production process to bring about M'. While M' presupposes surplus-labour as its ΔM component, it does not, however, presuppose M as the starting point of the circuit, since what emerges at the end point is not M' but C'. Since the production of M' directly does not require the conversion, prior to production, of M into C, such production is not dependent on the prior existence of capital.

Capital has not invented surplus-labour. Wherever a part of society possesses the monopoly of the means of production, the labourer, free or not free, must add to the working-time necessary for his own maintenance an extra working-time in order to produce the means of subsistence for the owners of the means of production... (Marx, 1983, p. 226).

And all forms of society apply labour-power to means of production. ΔC — or, more precisely in this case, ΔP (added product) — can therefore be increased only the more L P is exhorted (or improved) to produce more. That the surplus product will be transformed into surplus money is, however, dependent on the prior existence of money, else C' would not be posited as M'. Gold miners do not produce either money or the money-commodity, they produce the material gold which, if the economy in question is a money economy, is then immediately posited as money or the money-commodity. And as such it is posited as the *starting point* of the circuit of capital.

When a Marxist theory of gold, or the beginnings of such a theory has been offered, the circuit of gold-producing capital has usually been reproduced straight out of Marx's *Capital* and taken as axiomatic. Whatever discussion is offered amounts to a paraphrasing of Marx without the circuit itself being subjected to any scrutiny. We will show that Marx, whose concern was not with gold, but with capital, offers a circuit which explores gold *only in so far as it intersects with capital*. In order to correct this shifted spectrum, we shall discuss gold and examine capital in so far as it intersects with gold.

The circuit for gold production described by Marx in Vol. II of *Capital* is not wrong. It is simply weighted for an analysis of capital, rather than an analysis of gold. The circuit offered by Marx for gold production[4] is:

$$M - C \ldots P \ldots M'$$

This cannot be the *elementary* circuit of gold production, for it assumes capital as a prerequisite to gold production. In simple placer production, the simplest form of gold production — indeed, so simple as to be indistinguishable in its circuit from straightforward plunder[5]— we have simply:

$$\begin{matrix} MP \\ LP \end{matrix} \ldots P \ldots M'$$

But the conditionality of this circuit is already obscured by the presence of **M'**, a form of the product which is not present at the opening of the circuit. In other words, the product is already presented in a form in which it is insertable into a circuit outside of the economy to which its production belongs. But whence $\Delta \mathbf{M}$, or, differently put, how can $\Delta \mathbf{M}$ be known if **M** is not present at the opening of the circuit?

What looks like **M'** from the vantage point of money economy, looks, from the point of view of its primitive producer, as $\left(_{LP}^{MP}\right)'$. It is here that the special attribute of labour-power to produce more value than it itself has may be directly observed, for the circuit of gold production is now reduced to the elementary circuit of all production:

$$_{LP}^{MP} \ldots P \ldots \left(_{LP}^{MP}\right)'.$$

Where the product of labour-power does not assume the form of commodities, the circuit of gold production, or any production for that matter, stands as:

$$\mathbf{P}_{LP}^{MP} \ldots P \ldots \mathbf{P}'.$$

Production under conditions of barter presents the circuit as:

$$\mathbf{C}_{LP}^{MP} \ldots P \ldots \mathbf{C}',$$

Thus is it where gold is but a particular use-value, and not yet elevated to universal use-value. Exchange value has not yet settled into gold as its own particular corporeality. The production of gold is no different to the production of anything else. Its limits are the limits of consumption.

> It is, however, clear that in any given economic formation of society, where not the exchange value but the use-value of the product predominates, surplus-labour will be limited by a given set of wants which may be greater or less, and that here no boundless thirst for surplus-labour arises from the nature of the production itself (ibid.).

It is otherwise where gold has become money, for now the circuit becomes:

$$C_{LP}^{MP} \ldots P \ldots M'$$

Suddenly the production of an ordinary commodity transforms itself into the production of money — and that in the most direct way. *This* is the circuit of working to death, and not the one below, as suggested by Marx. Here slave-owner and feudal state alike set labour-power and means of production to the creation of that commodity which has come to embody abstract wealth, value-for-itself, and they do so *without capital*. Brute force, rather than money, is the opening element of this circuit, and it remains significant throughout. The opening of the circuit by money, M, presupposes generalised money relations, i.e., capital, which obviates (once up and running) the need for brute force as an element in its continued reproduction, thus also implying wage-labour. The presence of wage-labour and capital, rather than working to death, underlie the gold production circuit offered by Marx and which reads:

$$M - C_{LP}^{MP} \ldots P \ldots M' \quad \text{or} \quad M - C \ldots P \ldots M'$$

Whereas in capitalist production in general, $M - C \ldots P \ldots C' - M'$, production is *drawn into* the circuit of capital, gold production here *is* the circuit of capital. Under value conditions, gold production, even in its simplest form, viz., on the basis of the isolated individual free producer, truthfully expresses the circuit $M - C \ldots P \ldots M'$. The relations of capital are therefore already concentrated in him. Surplus-value production, the *conditio sine qua non* of capitalist production, expresses itself primitively in the free gold producer in that he is both capitalist and wage-labourer at the same time, thereby turning him into his own exploiter. Capital is not capital without the enhancement of M into M'. The efficiency of capital reduces itself to the smallest number of refluxes necessary to achieve a given enhancement of M' over M, or a given composite magnitude of ΔM. The gold producer, like any capitalist, achieves this by continually reducing the value of C. But, as has been explained above, there are both theoretical and objective limitations to such reduction. Ultimately, such reduction is confined to the value of the gold producer's own labour-power. He thus increases his surplus-value by reducing his own necessary value and this not through increased productivity, but through reduced consumption. The inherent tendency is for this to be done *without* resort to technological improvement. This is discussed under competition, below.

Where the gold producer works for others under pre-capitalist conditions, we get capitalist production without capitalist circulation. The labourer is not a wage-labourer and his labour-power, therefore, not a commodity. Surplus-value production does not proceed on the basis of exchange. In the same way as the product of such labour is more logically plundered than exchanged for, so is the labour which produces it more logically plundered than exchanged for. Historically this took the form of simply working the labourers to death, i.e., plundering their labour-power, and replacing them by plundering the living inhabitants of neighbouring societies as slaves.

> Hence in antiquity over-work becomes horrible only when the object is to obtain exchange value in its specific independent money-form; in the production of gold and silver. Compulsory working to death is here the recognised form of over-work (ibid.).

(Marx is here citing Siculus which means that the passage needs qualification, since Siculus was describing lode mining rather than placers, the latter of which set the then socially necessary labour-time for all gold production. Placer production, as suggested above, did not lend itself to production for others. Placer gold production, while near-bestial, did not amount to working to death. We could find no evidence to suggest that Marx knew of this distinction). Co-operative gold production involves a more sophisticated labour-process than that of the individual free producer. This becomes necessary where the gold has a higher value than the social average necessary for its production. Access to individual accumulation of money is here restricted by necessary labour-time becoming so high as to all but eliminate the surplus. Reduction of the value of labour-power to restore the surplus-product is impossible on the basis of free production, since no-one freely works himself to death. Introduction of technology and a technical division of labour is thus forced upon the gold producers.[6] While the technology remains relatively simple, i.e., largely producable by the gold producers themselves for themselves,[7] they remain in control of their newly-necessary means of production. But technology and co-operation notwithstanding, they are still producing the material of money and they will continue to do so while keeping their own personal consumption to the absolute minimum.[8] As soon as the new necessary technology begins to fall outside of the ability of labour-power to put in place (usually through its having to be brought in from outside), the co-operative begins to break down. Its members must then seek out those in possession of *capital*, and resume gold production on the basis of wage-labour, or abandon gold production altogether.

Where capital is generalised as the social basis of all production, there wage-labour is generalised as the social form of all labour. The form of economy based on capital distinguishes itself from other, earlier forms of economy in many ways. One of these is the fluidity and flexibility with which both capital (objectified labour) and (living) labour are applied to society's varied and varying productive tasks. All the particular instances of the allocation of capital and labour form part of a seamless landscape of general social resource allocation. The form of economy increasingly develops the facility with which social resources may be withdrawn from one particular allocation and directed at another. This may be split into any number of parts and become so many new allocations of capital and labour, or any number may be combined to form a new, larger allocation.

Where gold production proceeds on the basis of capital and generalised wage-labour, such capital and such labour become integral parts of the general mass of capital and the general mass of labour of that society. Their attributes are set by the general attributes of capital and labour for the society as a whole. The general tendency, therefore, is for gold-producing capital to approximate the average profitability of the economy as a whole, while for labour it is to approximate the average value of labour-power.

That the interface between, on the one hand, extractive industry in general and mining in particular, and on the other hand, industry in general is not entirely seamless, has already been discussed. But it has similarly been suggested that the general tendency is towards such seamlessness. This free flow of capital and labour, while embracing other industries first,[9] eventually also draws mining into its ambit.

But the value of labour-power in extractive industry in general is constrained by a different set of objective conditions, and in gold production in particular, is determined by a different set of laws, to that of industry in general.[10] The objective condition that mechanisation comes but tardily to mining, means that the constraint placed upon the value of gold by the fixed ratio at which it is represented by the token of value, can only be accommodated by keeping the value of labour-power in gold production low. But this low value of labour-power cannot be achieved on the basis of high productivity of labour, as in mechanised industry in general, for mining retains primitive labour processes for much longer. The difference between the two is in their respective standards of living. Low-value, low-productivity implies a low level of consumption, while low-value, high-productivity implies a high level of consumption. In the competition for labour, a higher standard of living for the same labour-time expenditure is always more attractive to the worker. The result is that now, for the sake of production of the money-commodity, gold-producing capital must become integral with capital in general, in order that it might have the wherewithal to, as rapidly as possible, mechanise its labour processes, while *at the same time* isolating its labour force from the labour force in general, in order that it might cling to its low-value, low-productivity labour force in an environment of low-value, high-productivity labour.

This simultaneous integration and segregation means that gold-producing capital is both part of capital in general and not part of capital in general. But not in the above sense alone. While production of all use-values is, ultimately, determined by consumption,[11] that of the money-commodity, being a universal use-value, is determined by the very existence of the economy which conjures it into being. Capitals compete for surplus-value through the disposal of their products on the market. Gold-producing capital does not do this. The effect which competition has of lowering the value of labour-power on the basis of increasing productivity of labour, therefore, does not apply to gold-producing capital. To further explore the circuit of gold production on the basis of capital, it is necessary to introduce the category of competition.

Competition and gold production

From the commodity's fixed price, the natural deduction is that there can be no competition between the various gold-producing capitals to dispose of the product: from the point of view of cost of production, the fixed price is kept *low*; from the point of view of competition, it is kept *high*. The implication of this is that: (i) the industry must dispose over a non-competitive mechanism for surplus-value apportioning; and (ii) the mechanism which drives *competing* capitals to constantly improve their productivity of labour therefore falls away. Does this mean that gold mines have no inherent drive to mechanise? Three writers would answer this in the affirmative, two basing themselves on the social role of gold (Williams, M., 1975 and Trewhela, P., 1986a), and another on an observation by Marx (Ticktin, H., 1991). Most, though, allege lack of mechanisation for different reasons. The implications of this lack of competition has not yet been seriously taken up on a theoretical level.

In Marx's explanation of how surplus-value is apportioned where production proceeds on the basis of capital, the role of competition is crucial. Capitalists dispose of their commodities at prices below their competitors' cost of production but above their own, thereby capturing the difference in value as their share of the total surplus-value in the market. It is, indeed, this totality of all capitals in competition with one another which, for Marx, defines competition:

> Conceptually, *competition* is nothing other than the *inner nature of capital*, its essential character, appearing in and realised as the reciprocal interaction of many capitals with one another, the inner tendency as external necessity. ...Capital exists and can only exist as many capitals, and its self-determination therefore appears as their reciprocal interaction with one another (Marx, 1973, p. 414, emph. orig.).

The question then becomes: What form does their competition take, and how do the individual goldmining capitals capture their share of the total surplus-value? Competition in goldmining takes two forms: (i) competition to buy labour-power and means of production; (ii) competition to secure equity capital. Given the fixed value at which the product is disposed of, such capital would, from the start, seek also to procure its inputs on a fixed-value basis. If prices fluctuate freely at one end of the circuit, then so, too, must they at the other, in order better to allow 'adaptation to changed circumstances' (which usually means 'passing on to the consumer'). However, where one end of the circuit has a fixed price item while at the other end price fluctuates freely, the tendency will be to attempt to fix the fluctuating end, rather than free the fixed end. The elimination of uncertainty answers the inner tendency of capital for proportional production, i.e., it gives capital greater control over its surplus-value production. Whether it can afford its investments and whether its returns are adequate need no longer be established *post factum* through the market, but can be known beforehand, working back from its fixed-value product.

Competition and the circuits of capital

Although modern goldmining capital is joint-stock capital, the processes we wish to examine may be more clearly illustrated by first looking at private capital. Wherever, along the circuit of capital a change of form occurs, there an interaction takes place with the total social surplus-value. It is at these points that a redistribution of surplus-value takes place. These are the points of competition — where capital has to influence that redistribution in its favour. Along the circuit of individual or private capital, this occurs in two places:

$$\underline{M-C} \ldots P \ldots C'-M'$$

Competition in M - C: Competition for inputs (means of production and labour-power). Here the object is to part with as little value as possible in exchange for a fixed value, C, required. Competition here takes the form of capital offering the lowest possible price above that offered its competitors. Competition also occurs at

$$M-C \ldots P \ldots \underline{C'-M'}$$

Competition in C' - M': Competition for surplus-value. The object is to gain as much surplus-value as possible for the available quantity of C. Competition entails demanding the highest possible price below that demanded by competitors. It is through competition in this part of the circuit that capital realises its share of the total surplus-value. Here the investor converts his capital back into a form which tells him whether or not his money has been successfully converted into capital.

That C' - M' takes place also implies its complement: M - C. A transformation of C' - M' for one capital is simultaneously a transformation of M - C for another. These two capitals are therefore use-values to each other, and their respective circuits are complementary. The form of competition here is that of one capital seeking to gain as much as possible for a fixed value (C' - M') while and the other seeks to part with as little as possible for a fixed value (M - C).

Alongside one productive capital seeking to carry out the metamorphosis C' - M', there are other productive capitals intent on the same metamorphosis. Failure to carry through this metamorphosis means failure as capital. These productive capitals are therefore in competition with each other to secure the complementary M - C. To any one capital seeking to effect C' - M', the presence of other capitals attempting the same is an unwelcome interference. It places a limit on how much surplus-value it can redistribute in its favour in the complementary metamorphosis C' - M'/M - C. This competition between productive capitals then assumes the form of each testing the other's capacity to meet the demand of the complement. Thus are they constrained to demand less than they otherwise might have. This double competition occurs at both ends of the circuit: at the opening M - C, and at the closing C' - M'.

The circuit of capital of a company run on a joint stock basis however, offers one more opportunity, in addition to those of the private company, for a redistribution of surplus-value, and hence, competition. Let us consider the circuit of productive joint stock capital:

$$M_I^S - C \ldots P \ldots C' - M_I^{S'}$$

(For the moment we need not yet distinguish the particular circuit of goldmining capital). Money (M) at the opening of the circuit originates, in this case, not only from an individual source, but also from a social source,[12] the stock market. The money thus has both a social component (M^S) and an individual component (M_I).

The first share capital is also attracted on the basis of an anticipated fixed portion of the surplus-value in the form of a dividend. But dividends can be withheld, reduced, paid in full or increased. Whichever occurs could either fully, partially or not at all reflect the amount of surplus-value produced. Flexibility and uncertainty as to the return on the capital invested is thus built-in. We return to this uncertainty in a moment.

Once the transformation C' - M' has occurred, M' again splits into its two original components of M^S and M_I, now enhanced as $M^{S'}$ and M_I'.

$$M_I^S - C \ldots P \ldots C' - (M_I' + M^{S'})$$

That portion of the surplus-value which would, in the case of private capital, constitute the profit, must here be divided between return to the individual money (M_I), and return to social money (M^S). Whether M' re-enters the circuit at M depends upon the magnitude of $M^{S'}$ in comparison to other magnitudes of $M^{S'}$ revealed on the stock market. Small dividends would tend to cause a redirection of capital (M) to those circuits yielding a higher $M^{S'}$. The controller of the enterprise therefore has to see to it that $M^{S'}$ is sufficiently large to ensure the re-entry of $M^{S'}$ into the circuit as M^S. If M_I' is greater than zero, one may assume that it will re-enter the circuit as if it were M' returning as M. The case of $M^{S'}$ is different. When the amount of surplus-value is relatively small (usually during the development phase of a company) in relation to the size of the subscription share value, it could occur that virtually the entire available surplus-value is given over to $M^{S'}$, resulting in Manchesterism. Should $M^{S'}$ be reduced even further or even withheld, difficulties with the re-entry of $M^{S'}$ as M may be encountered. Dividends are not paid and the share price drops, necessitating extraordinary measures, such as financing from internal sources. It is generally the companies which pay the highest dividends for the lowest subscriptions which attract the most new capital. This is reflected in an increase in the share price which in itself tends to attract even more capital.

Alongside the capital from which $M^{S'}$ has emerged at the end of a circuit, there are other capitals also seeking the return of $M^{S'}$ as M^S to start the circuit anew. Once $M^{S'}$ leaves the circuit it returns to the undifferentiated body of social capital whence it first emerged.

$$M^{S'} \rightarrow [\textit{social capital}] \rightarrow M^S$$

It is then available for re-extraction by *whichever* capital promises the greatest enhancement of M^S into $M^{S'}$ in the shortest possible time. The M_I which it joins to form a subsequent capital will not necessarily be that with which it formed the previous capital. A particular capital, therefore, has no guarantee of its existence from one circuit to the next, while, at the same time, capital in general becomes infinitely flexible.

This infinite flexibility is the positive side of social capital. The negative side is that of the uncertainty of the reproduction of a particular productive capital. Both of these attributes are reinforced by the fact that a more rapid circulation of capital can be had by avoiding production altogether, and simply trading in claims to future shares of surplus-value (the circuit M - M′ directly). In other words, in response to supply and demand for rights to portions of surplus-value still in the process of production, the magnitude of that very surplus-value can be altered. Although the circuit of productive joint stock capital may be given as:

$$M^S_I - C \ldots P \ldots C' - M^{S'}_{I'},$$

the entire capital is not concentrated at one point along the circuit, but at any one time is spread along the entire circuit. Part of it exists as money waiting to be transformed into commodity, another part has already been so transformed while another is locked into production. A further part lies in the warehouse as finished goods awaiting sale, while still another portion the firm must decide how it divides between what it keeps and what it puts back into social capital. Then, of course, there is the portion actually extant as social capital. Assuming a smooth circulation process, portions of the capital will regularly vacate one moment in the circuit for the next moment.

Any interruption in the circulation process means that only that portion of the capital existing in the form M′ (or $M^{S'}_{I'}$) will have been realised. The normal threat to the reproduction of productive capital, the failure to secure a complementary circuit at C′ - M′, and to a lesser extent M - C and during . . .P. . . , is here magnified by a dynamic beyond the bounds of its own competition and hence beyond its ability to influence, viz., speculation. M - M′, on a circuit-to-circuit basis, demonstrates the inadequacy of M - C . . .P. . .C′ - M′ to the concept of capital. The scope of this study does not include an exploration of that form of capital which accumulates through the commoditisation of capital itself.

While a joint stock company has some control (through the competitive process) over the size of its own dividends, it has no control over the price of its shares. What is more, these are often traded even *before* subscriptions have opened, meaning that the company loses control over even the issue price of its own shares.[13] In the early days of joint stock capital, prior to the regulation of the financial markets, vast amounts of productive capital, including goldmining capital, was laid waste through speculation.

There is no mechanism intrinsic to economics (or, more precisely, to capitalism) for solving the 'problem' of speculation. This has to be done from outside, by the state, on behalf of capital as a whole. Insofar as this problem has been brought under control, the contradiction of joint stock capital, of being simultaneously constrained and liberated, has been solved for goldmining capital by the lack of competition in the disposal of its product. This could be more easily illustrated if, for the moment, we confine ourselves to individual, as opposed to social, goldmining capital. The circuit of individual goldmining capital[14] may be given as:

$$M - C .. P ... M'$$

Formally (as well as formerly), the price of gold merely allocated a money-name to a given quantity of gold and facilitated the differentiation of national levels of labour productivity and hence the relative value of gold in the different economies. The 'sale' of gold was therefore but a formal requirement of exchange economy in a certain context, rather than a change of form of value from commodity into money, as in C - M. Since the material removed from the ground is *simultaneously* an ordinary commodity (a particular use-value) and a universal commodity, it does not require the metamorphosis C - M for its value to attain the form of money. It emerges out of the production process directly as money. Hence the absence of C ' in the notation of the production circuit.[15]

The absence of the closing transformation of the circuit means that the competition associated with it is similarly absent. No complementary capital is required and the circuit can be started anew immediately. At the opening of the circuit (M - C), though, goldmining capital finds itself in the same position as any other capital: it needs to compete for its inputs. It will attempt, like all other capitals, to reduce the competition associated with M - C as much as possible if it cannot actually eliminate it.

This applies as much to joint stock, as to individual capital. South African goldmining capital is no exception. But given the lack of incentive for monopoly at the closing end of the circuit, there is no such incentive at the opening end. The incentive is to co-operate in pursuit of scientific production and yet preserve their discretion as capitals. The individual companies *co-operate* to establish a body separate from themselves to serve as their *representative monopsony*. From its inception the individual mining companies have sought to co-operate in the area of especially labour procurement, but also in all other factors of production. And for this purpose was the Chamber of Mines established.

But the real lessons to be derived from South African goldmining capital, lie in its nature as joint stock capital. The circuit of joint stock goldmining capital may be given as:

$$M^S_I - C \ldots P \ldots M^{S'}_{I'}$$

As has been suggested above, the crucial distinction between joint stock capital and private capital lies *outside* of the circuit of capital, in the undifferentiated body of social capital. What is crucial is not what lies between $M^S_I - M^{S'}_{I'}$, but between $M^S \rightarrow M$, between the close of one circuit and the opening of another, for an objective break is now reintroduced. This break describes not a change in the value form (as in $C' - M'$), but in the capital form. Capital must here pass from the particular to the general and back to the particular. While it sojourns in the general, two conditions obtain: the circuit of which it had formed a part is in danger of not being reproduced; and, it is now free to enter any other circuit it chooses.

The absence of $C' - M'$ means that the second interface between particular capital and general social capital is brought right back into the circuit to coincide with the first interface, $M - C$. The entire production process, therefore, already lies within the domain of social capital. The attribute of particular capital, that of vulnerability to abandonment is largely banished from the circuit, and the attribute of social capital, that of infinite flexibility and applicability, is extended into and over almost the entire circuit of goldmining capital. A contradiction, nevertheless, persists, in that the attributes of social capital are brought into the circuit of what remains *particular* capital. The contradiction must express itself by periodic attempts on the part of particular capital to reassert its control over its own circuit against that of capital in general. This is clearly demonstrated in the early history of South African goldmining, both in attempts at collusion, and in the control structure of the Groups. Part III is given over to an examination of this.

Williams' theory of gold: a critique

The beginnings of a political economy of gold, based on a theory of luxuries, rather than a theory of money, is offered by Williams. Drawing on Marx's discussion of the cost of circulation in Volume II of *Capital* and of luxury production in Part III of *Theories of*

Surplus Value (see Marx, 1977a, pp. 138-9 and 1972, pp. 245-52, 349-50). Williams *fuses* these two discussions in order to present gold (and diamonds) as luxuries.

In order to show that, 'the goldmining industry is unable to produce surplus-value in its specifically relative form' (Williams, M., 1975, p. 6), Williams has to make money assume some of the attributes of luxuries, and luxuries some of the attributes of money. When Marx observes, in respect of *monetary* circulation, that, 'These commodities performing the function of money enter into neither individual nor productive consumption' (Marx, 1977a, p. 139), Williams deduces that the goldmining industry (and now he quotes Marx on *luxury production*), 'cannot produce any relative surplus-value and, in general, cannot produce *that form* of surplus-value which results from the *growing productivity* of labour *as such*' (quoted in op. cit., p6. The emphasis is Marx's and is from 1972, p. 350). He dismisses increases in the organic composition of goldmining capital as having not, 'the slightest influence on wages, on the *value* of labour-power', (again quoting Marx on luxuries) the reason for this being, 'since gold, in its capacity as money does not enter, either directly or indirectly, into the consumption of the workers'.

This is where Williams' creativity becomes its own undoing. Consumption, in contradistinction to production, is conflated with necessary-value consumption of labour-power in contradistinction to surplus-value consumption of the non-producers. When Marx speaks of money entering into, 'neither individual nor productive consumption', it simply means that the commodities which form money are not available for consumption *while they remain in circulation*. Differentiation *within* consumption is not made. Such commodities must, in the first instance, be articles for consumption else: (i) they would never become money; and (ii) the point becomes a tautology.[16] Marx's statement is not as emphatic as its rendering by Williams suggests when he says that the money-material 'in no way' enters into consumption.

In the case of luxuries, Marx distinguishes between consumption by the wage-labourers, and into which luxuries do not enter, and consumption by those who do consume luxuries. Here the distinction between consumption and circulation is not at issue. The implications of the conflation by Williams are: (i) luxury items constitute money; (ii) the material of money, e.g., gold, enters into the direct consumption of the non-producers.[17]

The discussion of the peculiarities of surplus-value production in the luxury industries becomes directly transferable to that industry which produces the commodity which serves as money. Williams is now in a position to deduce that the goldmining industry, therefore, 'cannot produce any relative surplus-value and, in general, cannot produce *that form* of surplus-value which results from the *growing productivity* of labour *as such*'. The productivity of labour in gold-production, Williams concludes, cannot have, 'the slightest influence on wages, on the value of labour-power', since gold, 'does not enter into the consumption of the workers'.

Williams' associating the money-commodity with luxuries is entirely arbitrary. Our own theory stands as critique of it. But the weakness of Williams' theory is even empirically obvious. What are the workers' wedding rings and tooth-fillings made of?[18] Consumption, in its most elementary natural determination, means eating. At one point in early South African goldmining history, the mines made advance payments *in cattle* to entice especially Pondo tribesmen to the mines. This particular form of money had the advantage that after it has been excessively consumed, that fact registers in the distension of the belly. A wealthy man, therefore, was a man with a big stomach.[19] It has already been shown that an increase in the productivity of gold-producing labour has a very direct effect on the value of labour-power. In conclusion Williams claims that, 'It is for this reason, Marx notes, that the producer of gold will seek, in practice, "to depress wages of labour below its value, below its minimum"', when Marx notes nothing of the kind. Marx refers to the producer of luxuries, and not to the producer of the money-commodity. This book has argued that the determination of the value of gold-producing labour-power expresses a relationship between gold as use-value, gold as money and gold as capital. The question of luxuries does not enter into that determination.

Williams' theory reaches its nadir when his inadequate study of money leads him to conclude from Marx's observation about the circulating medium condemning 'a part of the social wealth' to 'assume this unproductive form', that, 'this means that the diamond and gold industries in South Africa must secure their necessary materials from the surplus-products of Departments I and IIa... without their products entering into the production process of either' (op. cit., p. 8). One needs merely point to the existence of glass cutters and diamond-tipped drills, or electronics and insulating glass, to empirically show Williams' weakness. 'Regarding the production of luxuries', says Williams, ' — and this would apply to the diamond industry in South Africa — Marx emphasises', (and now he quotes Marx directly) 'their [the workers'] product, in the form in which it exists, cannot be transformed into capital, either constant or variable capital'. Not only is this wrong because Marx is referring to luxury items, but also because gold-producing workers (in the period in question) had a product which was immediately in the form of their wages, i.e., variable capital — not only their own, but that of all workers. Diamond-digging workers have a product which forms an important part of the constant capital of many industries, not to mention mining itself. Not only does Williams have a weak theory, he clearly has not done a proper empirical investigation.

Notes

1. This is briefly sketched by Marx in the *Grundrisse*. (see 1973, pp. 150-1).
2. This is touched on again briefly in Section II.
3. The following discussion is based on Marx, 1983, Ch. IV and Marx, 1977a, Chs I-IV and pp. 330-43.
4. The circuits described below are based on those as laid out in Marx, 1977a, Ch 1: C = Commodity(ies); M = Money or Capital; $\ldots P \ldots$ = The production process; -' (prime) = enhanced/of heightened value, e.g., M' being M enhanced to a higher value, etc.; MP = Means of production; LP = Labour-power; M_I = Individual (private) money; M^S = Social (anonymous) money; A dash (-) indicates that a change of form takes place, e.g., $M - C$ = Transformation of value from the money form into the commodity form.
5. '...the wealth which these brigands [the Cossacks] had obtained through plunder and placer mining' (Del Mar, op. cit., p. 377).
6. Referring to Tacquah (pre-colonial West Africa), Lock estimates that, 'Taking a fair average, it may be said that cutting out will take occupy one day, crushing a second, and washing two more - four days in all; the return may be 3 dwt., to be divided amongst 4 miners and 4 washers' (Lock, op. cit., p34).
7. and the land tenure system allows it.
8. 'Some 4 or 5 miners were working in 3 of these mines, but only extracting fragments of ore from old shafts and levels; these ores held 7 to 10 oz. of fine gold per ton, but the quantity is so

insignificant, that it can only afford a miserable existence to such men' (ibid., p. 200).
9   Agriculture, too, does not easily submit to the free flow of capital and labour.
10 This has been discussed in the Chapter: *Gold as Money*.
11 Even though, in capitalist economy, it might take the *form* of production for exchange.
12 It may be argued that a bank is a social, rather than an individual (private) source of capital and that, hence, where M consists, either wholly or in part, of a bank loan it should be considered along with joint stock capital. The difference, though, is that the bank enters into a fixed relation with the borrower of capital and expects a fixed share of the surplus-value for its advance. As soon as the bank's capital and its share of the surplus-value have been handed over, the relation dissolves. That portion of the circuit which flowed through the bank closes and, regardless of how much surplus-value has been produced, has to be negotiated anew in order to form part of M again. Bank capital can, therefore, be subsumed under individual capital and be denoted as simply M. While the bank capital formed part of M, there was no reason intrinsic to the circuit for why M' would not re-enter the circuit as M to start it anew.
13 In the City of London this is known as 'the grey market'.
14 Based on Marx, K., 1977a, p. 331).
15 Strictly speaking, this should read $M-C...P...C'/M'$, but to express it thus at this point would serve no explanatory purpose.
16 In societies where more directly consumable articles serve as the money-commodity, such articles are usually withdrawn from circulation and consumed before their use-value as ordinary commodities expires. It was prime cattle, and not old and sickly beasts, which represented abstract wealth in Xhosa society.
17 Williams does appreciate that problems of this kind do underlie his theory, which is, perhaps, why the following extraordinary statement appears in a footnote:

> Although the money-material and luxuries have a number of characteristics in common, ...this does not mean that they always share the same characteristics. On the

contrary, in its capacity as the money-material gold confronts the entire world of ordinary commodities, including luxuries, as the one exclusive commodity with the properties of money. This makes it all the more necessary when dealing with capitalist development in South Africa, to pay careful attention to the roles played by the different commodities in the reproduction process. Without making such rigorous distinction much in our analysis would remain blurred and obscured (op. cit., p. 33n10).

18 Amongst a section of the working class in Cape Town, it is considered beautiful to have slits of gold between the teeth. Sometimes perfectly healthy teeth are even capped in gold.

19 In plays put on by unionised African workers in South Africa, the boss is often caricatured as a man with an outrageously large paunch.

Part Two

GOLD PRODUCTION, CAPITAL AND THE VALUE REVOLUTION OF THE 19TH CENTURY

Introduction

The value of a commodity is the amount of labour-time necessary, on the average, for its production. This holds equally for labour-power and for other commodities. The value of commodities rise and fall with the fall and rise of the productivity of labour. It is possible for the productivity of labour in the production of just one commodity to rise while that in all others remains the same. It is also possible for the productivity of labour to rise for some commodities while it drops for others, and that the degree of rise or fall can vary from commodity to commodity. When a major change in the productivity with which the money-commodity is produced came about, without any change in the general productivity of labour, then a revolution in the *standard of price* has occurred. This happened, e.g., when the commercial economies were flooded with plundered gold from the New World. However, when such a rise or fall in labour-productivity is generalised across the economy, we may speak of a value revolution. The so-called Second Industrial Revolution, which started around 1850 and lasted until the turn of the century, may be characterised as such a revolution. No commodity, including labour-power and the money-commodity, escaped its influence. And it is the influence on the latter two which interests us in Part II of this study.

4 The changeover from placers to lodes

The technological revolutions of the nineteenth century affected gold production in three distinct and critical ways. The first was in the development of communications and transport. The invention of the telegraph, steamship and railway allowed, firstly, for news of a gold strike to be disseminated much more rapidly and widely than had ever been possible before, and, secondly, for gold seekers to rush in more speedily, in greater numbers and from further afield than had previously been the case.

Gold rushes have occurred throughout history, but prior to the period under review, these tended to be local affairs. The *Encyclopædia Britannica* identifies three great eras of expansion in gold production. The first was the exploitation of the gold of the Americas (both by mining and by plunder); the second, being the period 1850-1875, which includes the Californian and Australian rushes; and the third is the period 1890-1915, which includes both the last great placer rush (Klondike) and the first setting of goldmining onto a scientific basis (Witwatersrand) (Encyclopædia Britannica, 1982, Vol. VIII, p. 237). The rush to exploit American gold after Columbus' explorations can hardly be described as a 'local affair', but we exclude this from our definition of a gold rush since extracting this gold remained outside the reach of the single individual. The mass of people who would otherwise have formed the *dramatis personæ* of a gold rush had an insurmountable barrier between themselves and the gold, the Atlantic Ocean.

By the third period (1890-1915), one is talking as much of a rush of people, as of a rush of capital. But already by the time of the California gold rush, 'rushing' involved an enormous capital outlay. Gold seekers came from all over the world, many having sold everything they had and spent their life savings to get there. The gold digger had every incentive to, at the very least, recoup his investment. Goldmining was beginning to hear the distant roll of capital. The storm was to break in the 1890s.

During the Middle Ages, the principal source of new gold for Europe was known to be west Africa — the gold being shipped north across the Sahara and exchanged in the markets of the Maghreb[1] — yet there was no flood of fortune seekers to west Africa. It would be interesting to explore how general social relations of personal bondage, as opposed to free individuals, were able to respond to the known presence of a source of treasure within geographical reach. The exploits of the *Conquistadors* may be said to stand as a precursor of the nineteenth-century-type gold rush.

The literature tends to be somewhat overawed at the scale and intensity of the nineteenth century gold rushes (see, e.g., Richardson, P. and Van Helten, J-J., 1984; Green, 1968 and 1985 or Fordyce, W., 1924). From the nineteenth century[2] on, gold sources were inundated with people from around the globe almost as soon as knowledge of them became public. Assuming no revolution in goldmining technology — and none had yet occurred by the time of the Californian and Australian discoveries — the sheer volume of people alone would bring about two unprecedented developments: (i) a vast increase in world gold supply from *production*, rather than from plunder; and, (ii) a rapid depletion of the sources.

The depletion of the placers

Up to the 1850s, whether from placer or lode sources, gold was still extracted everywhere, as in Georgia, 'mostly by rude and primitive appliances' (Lock, op. cit., p. 171). The most sophisticated of these were in placer mining: the long-Tom, the sluice and the riffle box represented the height of placer mining technology. The labour-process of placer mining continued around the world much as follows: 'The miner loosened the auriferous soil with a pick, and shovelled it into a pan from which, by skilful manipulation in water, the clay, sand, and pebbles were removed, and the heavier particles of gold-dust left behind' (ibid., p. 125). The first machinery to appear in surface work was confined to crushing hard ores, later to be extended to metal extraction. The crudity of these devices, even as late as the 1860s, may be gleaned from the following description of central American mining:

> The silver and gold ores are crushed in a basin of masonry, in which rises a vertical shaft, driven generally by a horizontal water-wheel. This shaft has two arms, from each of which is suspended a large stone or boulder. These are the crushers. After

the ore is reduced to sufficient fineness, the metal is separated by mercury: 'a long and expensive process, which is now beginning to be facilitated and cheapened by the introduction of the German or "barrel process". The machines for crushing the ore have, however, as yet, undergone but slight improvement' (ibid., p. 121).[3]

World gold production stood at an annual average of 650,000 ozs. for the decade 1831-40. The years 1849 and 1850 witnessed the Californian and Australian discoveries. Average annual world production for the decade 1851-60 stood at 6,300,000 ozs., which then gradually fell back to 5,200,000 ozs. between 1881 and 1890 (Vilar, P., 1976, p. 328). An examination of the output figures for these two areas will show that production reached its peak very shortly after the fields were declared, then tapered off to eventual depletion. In former times, goldmining provided a more steady output with a less volatile effect upon the economy in general. The drop in average annual output from the decade 1861-1870 to the decade 1871-1880 was greater than the average annual production for the whole period 1781-1830. (See Table 1, below).

Table 1
Total world precious metals production, 1741-1910
x 1,000 ounces

Period	Gold	Silver
1741-1760	791	17,100
1761-1780	665	21,000
1781-1800	572	28,300
1801-1810	572	28,700
1811-1820	368	17,400
1821-1830	457	14,800
1831-1840	652	19,200
1841-1850	1,762	25,000
1851-1860	6,313	26,500
1861-1870	6,108	39,000
1871-1880	5,472	66,800
1881-1890	5,200	97,200
1891-1900	10,165	161,400
1900-1910	18,279	182,600

Source: Vilar, P., 1976, p331.

The experience of Alaska and the Klondike illustrates the point even more strongly. The new placer-mining technology which was being developed during the course of the exploitation of the Californian and Australian placers, was applied in full force to Alaska and the Klondike from the start. Alaskan gold, discovered in 1898, reached peak output between 1900 and 1906, then dropped off very dramatically. The extraordinarily rich Klondike, delivering more than half an ounce of gold to the pan, was opened in 1896, peaked in 1900 with 1,350,000 ozs. and then dropped, forcing a changeover to lode mining. The presence of cutting-edge technology does not change the basic nature of placer mining. 30,000 prospectors, having risked everything to cross an Arctic mountain range, descended upon Dawson City, where, 'absurdly primitive individual prospecting went on side by side with fairly modern establishments' (ibid., p. 328).

If we consider placer mining from the time of the opening of the California fields in 1849 to the turn of the century, then this type of mining has itself undergone a number of changes. These relate to the depletion of one variety of placers (shallow placers), and the changeover to another variety (deep placers). When we said earlier that the placers had set the average socially necessary labour-time for gold production through the ages, this was still somewhat broad. It was the *shallow* placers which served as the standard for all gold production. In the production of gold from shallow placers, gold presents itself as a measure of value in the most direct way. It is immediately obvious how much gold was produced by one labourer in a given time because this gold was producable by the single individual producer. Essentially, no co-operation was necessary and no further processing of the material was necessary, while production could proceed on the basis of the meagre aids which a miner could carry on his back. Thus was by far the greater portion of the world's gold won for all time prior to the developments discussed here.

It must be kept in mind that in goldmining, as in all extractive industry, the substance extracted is not renewed. Every location holds only a finite quantity of gold, so that for every ounce produced, an ounce less remains. When a goldfield is inundated with independent producers, each producer must attempt to secure for himself access to as great a portion of that fixed quantity as possible. This may be done in one of two ways: (i) by claiming *and holding* a larger portion of the ground area to be mined; or (ii) by processing larger quantities of ore in a given time.

This second alternative does not in and of itself lend access to more gold, but does allow the miner to more rapidly exhaust his claim and move on to another. By the time he abandons a claim, he must be fairly certain that no (extractable) gold remains. But the one-man digger can never be certain whether his next panfull of dirt contains Nirvana, or whether he should abandon the claim and start on another. By producing faster, he reaches a point where he comes up against the limits of his own necessary labour-time sooner than do other diggers around him, all else being equal. He is thus both forced and enabled to move on to another claim. His own necessary labour-time is, therefore, both a blessing and a curse to him. But this is only the objective limit. It must not be forgotten that this production demonstrates the equivalent of greed in consumption and accumulation in circulation. The moment a free miner's output drops below the average for the area, abandoning his claim and starting a new one presents itself as more efficacious. Hence do we read in reference to,

Lachlan District [New South Wales]—Warden Dalton, in the Annual Report of the Department of Mines for 1877, refers to the neglected deep leads on the Lachlan gold-fields, around Forbes, the gutters of the several leads having been hastily worked to the deep and wet ground, and there abandoned. The smallness of the claims prevented the application of the machinery and plant necessary to enable the holders to contend with water, quicksands and swelling schists with any prospect of success. With such imperfect appliances as were used in shallow and sound ground, the miners had not thought of reef-wash or benches, and when they had broken into the adjoining claim and sent up the last bucket of wash from the gutter, they abandoned the holding. Besides miners in those days, who knew that they had wash before them that would yield $1/2$ oz. to 1 oz. of gold to the bucket, would not willingly lose time by sending stuff to the surface that would not return more than 1 oz. to the load. In every instance, there was a strenuous effort made by each party to get their share of the gutter worked out before that of the adjoining shareholders, fearing that if these men were worked out, and abandoned the lead before they had secured the best of their wash-stuff, they would lose it, as they knew that single-handed (*sic*) they could not cope with the water and other obstacles that overlay the auriferous drift. These difficulties, in an aggravated form, present themselves to small co-operative companies without capital at the present day. The

> leads are not exhausted—the reef-wash, benches, and tributaries are still virgin ground... Knowing all this, men hesitate to face ordinary mining difficulties that are of common occurrence in other parts of the world, and are invariably overcome by skilful engineers and the application of capital...
> ...Not a single lead has been traced to its termination, not farther than could be done by unassisted manual labour (Lock, op. cit., pp. 511-3).

Under the influence of such competition,[4] technology to make placer mining more efficient began to be introduced. These operated on the same principle as the pan, indeed, were mere developments out of it. They all involved the movement of ore across various obstructions by means of water through a controlled channel. Deeper down the water would flow more slowly than closer to the surface. The heavier particles, gold being one of the heaviest, would only sluggishly be conveyed by the slow-moving water and thus be trapped in the obstructions.

> The first improvement on the pan was the rocker; afterwards the 'Tom' was introduced from Georgia, and with it the sluice. In 1850 and 1851, the two latter devices began to be generally employed, in consequence of the greater convenience of water-supply afforded by the mining ditches. The first ditch of importance was made in 1850; and its success so stimulated imitation that in the course of 8 years, 6,000 miles of mining canals had been constructed, at a cost of more than $15,000,000 (£3,000,000), in California alone (ibid., p. 125).

While some of the new devices, such as the cradle (also called the box) could be operated by one man, the others involved some kind of co-operation between the miners. Not only was a certain amount of capital now required (small as this might have been when compared to later requirements), but to operate most of them called for the co-operation of at least two, usually five or six, and sometimes more men, working as a team.

New social relations of production began to emerge. Sometimes these teams were partners working co-operatively, at other times we see the beginnings of a capital-labour relation in goldmining. In the latter case the earliest form appears to be one where one miner owns the equipment while five or six others, operating the equipment *alongside him*, do so for a wage. Where a large number of claims was worked as one, this tended to be by a constituted company, either private or joint stock. At the same time the individual gold-digger persisted. It is conceivable that these relations were fairly fluid, with easy interchange for many between the state of lone gold digger and the state of wage-labourer.

Naturally, the shallow placers, which declare their presence to all passers-by, were the first to be discovered and the first to be depleted. but the exploitation of deep placers, too, have a long history. The mines in the mountains of ancient Egypt were of this variety. And that was the point about them. Deep mining could proceed in mountains well before the development of deep mining technology, because the rockwalls of the mountain itself serve as natural props and shuttering. The depletion of the shallow placers forced a closer concentration on the mining of deep placers. But while, consistent with placers, the gold was native and quite discrete from its matrix, there were now the additional problems normally associated with conventional mining: uncertainty as to the location, extent and quality of the ore; the inaccessibility of the ore; the need for equipment greater in quantity and complexity than that commanded by the one-man digger; the need for capital; and the relative growth of prospecting as an integral part of mining operations.

Access to the deep placers was effected in two ways: (i) the digging of shafts, inclines or adits and their development into galleries and stopes, as in conventional mining. This was the first response to the depletion of the shallow placers. Now, within placer production itself, the value of gold was changing, as more labour now had to be expended on the production of a unit of gold than would be the case with shallow placers. '[As] ...the first gold discoveries were virtually free, cost of production therefore increased rapidly. The impact on the ratio of the values of gold and commodities was instantaneous, while the more general effects made themselves felt gradually'.[5] (ii) Deep placers were also responded to by the washing away of overburden by a technique called *ground sluicing* or *hydraulicing*. This particularly destructive technique involved the application of a powerful jet of water to the walls of a valley. Thus literally scouring out the entire soil cover of the valley and washing it down through a control point to an area located at a lower level and capable of receiving the tailings. Gold was captured at the control point by the washings being directed through a system of sluices. By 1877, one observer was able to declare that, 'The placer mines of Goochland county are now successfully worked by the Hydraulic process, which is so extensively and profitably carried on in California' (quoted in ibid., p. 185).

Neither deep mining (deep digging), nor ground sluicing could proceed without the investment of relatively large amounts of capital. As well as capital, large amounts of labour-power now had to be expended before the first ounce of gold saw the light of day. Wage-labour now became the dominant form of labour, while the joint-stock company became the dominant form of capital. There was still room, though, for the private company. The social relations of gold production became a great deal more complex, and the connection between the value of gold and its role as measure of value was no longer immediate. Not only was the ruler re-calibrated, but the calibrations were fuzzy. In the case of the one-man miner, the cradle introduced a simple *decline* in the value of gold. But at the same time, devices such as the sluice and the long-Tom, made the value-relation more complex. By the introduction of ground sluicing, a handful of gold dust produced no longer indicates to the miner whether or not he will reproduce his labour-power. The capital-labour relation removes him from his product. While he will continue to be exhorted to produce more — though now by capital rather than by money — he will never know the extent to which he has exceeded his necessary labour-time requirements. The emergence of wage-labour in goldmining, therefore, itself obscures the value of gold. Gold being the measure of value of all other commodities, it would appear that one may here speak of the secularisation of value.

But most important of all, is the depletion of placer deposits, and the technical and financial inability of the mining industry to make the transition to quartz reef or lode mining. It was in California and Virginia in the 1870s and 1880s and in Australia in the 1880s, that this threshold was first reached, and in its South African theatre in the 1890s, for unique local reasons, that this threshold was first systematically broken through. The rapid diffusion of technology that came to characterise the nineteenth century played a large part in the simultaneous depletion of placers around the world. According to Lock,

> The enormous, easily-worked alluvial deposits of Australia, California, and Siberia, have been exhausted over large areas, though many similar deposits must yet remain undiscovered. In the more difficult operations of extracting gold from mineral lodes and complex ores, reliable evidence from all parts of the world shows that most of the processes at present in use, or the

methods of carrying them out, are far from satisfactory, as they entail the loss, on the average, of one quarter to one third of the gold present in the material operated upon (ibid., p. v.);

In the mountains between Pretoria and Potchefstroom, the gold hitherto found has been, for the most part, in quartz, and cannot be extracted without more machinery and cheaper fuel than the miners have as yet been able to command (ibid., p. 23);

It [the mines in Gavilanes, Mexico] ...was found to produce very rich ores from the surface to the depth of 60 varas (167 ft.), where a kind of black ore was discovered, which resisted all attempts to reduce it to advantage. By smelting, it yielded little or nothing, and by amalgamation, although the quantity of silver produced was very considerable, the loss of quicksilver was so great as to leave no profit (ibid. p. 107);[6]

These lodes ...at Major's Creek [New South Wales] ...have been proved to contain a large quantity of gold. So far the only process by which the gold can be extracted from these lodes is a chlorine process in use in Victoria but the cost of carriage of the material to the works is so great that the margin left barely covers the cost of raising the stone (ibid., p. 486).[7]

This implies that goldmining would henceforth have to be pursued on a vast scale, and already by 1865, when the Tati-field was being rushed, 'several joint-stock companies and private associations were formed to test its richness' (ibid., p. 16), thereby relegating the heyday of the rough-and-ready one man gold digger to the annals of romantic nostalgia. Placer, and especially alluvial gold had a certain certainty about it. The miner always knew where to look and had a fairly good idea of the general extent of the source. The level of technical development at the time the placers started running out (in the 1870s) was such that detection of a lode or reef could only be done by sight. The thickness of an underground ore body, its yield, its direction and continuation could not be known unless it was actually physically exposed. Even test drilling for samples was a more or less hit-and-miss operation (Eaton, L., 1948, p. 43). The investment of huge amounts of capital just to get to the lode therefore had considerable risk attached to it. Goldmining would, henceforth, require much more than the mere value of labour-power of the actual gold miners.[8] Hence Lock's concern:

> Much will depend upon the systematic development of such reefs. Their variable thickness and extent suggest the precaution of providing in many instances a reserve fund for occasional prospecting, and the advantage of the management of the mining operations being under interested and local directors. These and other considerations should not be disregarded, for by mis-management, either of a company's mine or of its funds, a good mine may be brought to a standstill, many incautious investors ruined, and the development of the mining industry retarded. Most of the mines are at present held and satisfactorily worked by *bonâ fide* miners, aided sometimes by others who are represented by waged men. In such private companies as these, where the labour and capital employed are mutually interested, the first development of a mine is generally attended by the most satisfactory results (ibid.).

By 1876, even what remained of the placers had begun to reach the limits of the most advanced placer technology. Its problems, once the ore was 'at grass', were the same as that of lode mining. These relate to extracting the gold from the ore. So fine and light had the gold particles become that the age-old method of panning, or any variation on it, would simply wash the gold away. An observer at a mine in Virginia saw the problem thus: 'Should this [the Lightning Amalgamator] process prove a success at the Bankroft Mine, it will at once create a revolution in goldmining in the South, as well as elsewhere. Low grade ores can be worked at a profit, where with any other method, they would not pay expenses' (Pollard to Lock, quoted in ibid., pp. 186-7).He says further that,

> Much of the gold in Virginia is very fine, and cannot be saved by amalgamation. Wet crushing is used generally. What we want is a better saving process than yet known to me, unless chlorination shall serve that purpose. To my mind its chief objection is that it is expensive. I should also advocate dry crushing. Upon a good many properties the gold is lamellated, and so thin and delicate that it will be readily carried off by water (ibid., p. 188).

This is the situation which confronted the goldmining world from around 1870. The depletion of the placers does not mean that individuals will never again be able to dig for gold. Where gold accessible to human hands will again be discovered, there will human hands again dig for gold.⁹ Mining was now more than ready for the industrial revolution.

Notes

1. 'The Venetian traveller Cadamosto, who visited Western Africa about 1454, alludes to the gold-field of Timbuctoo..., and declares that the gold coinage of Portugal, Spain and Italy, in the 14th and 15th centuries, was entirely supported by supplies from this region. Most of the gold then exported from the Soudan [contemporary western sahel region— FS] would seem to have found its way by the slave and ivory caravans to the Mediterranean ports of Tunis, Fez and Morocco. The large unwalled town of Gyni, in the kingdom of Melli, is specified as a headquarters for the exchange of gold and salt' (ibid., p. 37).
2. Although Russia was the world's principal gold producer from 1823 to 1837, its contribution, from the point of view of political economy, belongs properly to the pre-industrial era. On these grounds Russia is excluded from this study. Others have grouped Russia along with California, Australia and South Africa, for the simple reason that they were all important in the nineteenth century (see, e.g., Richardson, P. and Van Helten, J-J., 1982).
3. An attempt was made in the 1860s to introduce a traction engine to the Tati field.
4. Competition in goldmining takes a form very different to that in other areas of production, given the lack of competition in disposal of the product. Competition here manifests itself as rivalry in a contest against nature (see Sections I and III).
5. Vilar, op. cit., p. 326. 'It may safely be held that no gold has yet been produced in Alaska or the Klondike at a lower wage than $10 and much at $15 to $20 a day. In some instances $40 a day were paid' (Del Mar, op. cit., p. 443). 'For every dollar extracted from the sands of the extreme North more than a dollar has been expended' (ibid. p. 443n3).
6. This was, in fact, a gold mine. Mexican gold almost always occurred alloyed with silver, necessitating some separation process. Thus, despite the 'considerable' amount of silver

retrieved, its object, liberation of the gold, had not been attained, hence the failure.

7 'Up to 1871, only alluvial washings were carried on, and the wealth lying in quartz reefs was comparatively neglected' (ibid., p. 479).

8 The geologist thus becomes a highly-prized specialist, whose 'considerations are of importance to the miner, as showing that the thickness of the reefs cannot be relied upon for continuance in either length or depth, and that some reefs may even thin out altogether', Lock, op. cit., p. 497. Only large mining companies could employ geologists.

9 The name Sierra Palada will probably forever be associated with grotesque images of human ants jostling one another for a toe-hold as they *kriewel* their way up and down vertical mud cliffs, toying with death. In Bourkina Fasso, the Sahara swallows many a gold digger who dares to dislodge one grain of sand too many.

5 Universal industrialisation and social capital

Much has been written on the so-called 'second industrial revolution' of the latter half of the nineteenth century (see, e.g., Chandler, A., 1990, Clapham, J., 1961, Dunning, J., 1988, Kemp, T., 1969, Landes, D., 1981, Payne, P., 1967, Pollard, S., 1988, Trebilcock, C., 1986, Trewhela, P., 1970). The real significance of this series of developments lies neither in the unquestioned expansion in the scale and concentration of production, nor in the vastly increased mechanisation[1] and improved productivity of labour,[2] but in the revolution in the capacity of society to allocate its resources. Up to the first industrial revolution, resources, whether in the form of objectified or of living labour, were *locus specific* in their application. Particular capital (in the widest sense) engaged particular labour at a particular site for the production of particular products. The conscious intention, both on the part of such capital and on the part of such labour, was to remain so engaged by each other at that place producing that product. Both labour and capital thus developed attributes particular to the particular locus, differing from those of capital and labour in other loci. Levels of skill, earnings, productivity, intensity of labour, profitability, etc., were all locus specific.

This locus specificity is, of course, another way of saying 'insularity'. This insularity was preserved over centuries in that wealth (whether the estate, the farm or the firm) was handed down from father to son; and skill (in the form of a particular artisanal trade), similarly, was handed down in this way. In this way a particular capital was associated with a particular labour for generations without either interruption or dilution.

If the historic significance of the first industrial revolution was to break the bounds of particularity, the role of the second was to formalise universality. 'South Sea Bubble', *société anonyme*, the stock exchange, multinational corporation, *Wertpapier*, currency speculation, etc. are all terms illustrating the freedom of capital from particular applications — a development which rose to full prominence in the latter half of the nineteenth century. The specific mechanism which gives effect to this universality of capital is the rise of social capital, i.e., the commoditisation of capital. Capital itself becomes an object of capital: it is bought and sold on the stock exchange.

Labour is similarly freed from particular applications by its commoditisation, i.e., its transformation into wage-labour. It becomes effectively universalised by the increasing convergence of the attributes of all labour towards a social average. This tendency is, of course, countered by the tendency of industry to specialise, having the effect of partially shifting the common attributes of labour out of production, and into the general social environment. Thus does a definite level of literacy, numeracy, punctuality, cleanliness and health, and technical proficiency become associated with labour in general, rather than with any particular group of workers. Convergence towards an average occurs more rapidly for some industries than for others, depending partially on the peculiarities of production, e.g., the extent to which mechanisation is possible, the peculiarities of capital, e.g., a predisposition to keep money 'in the family', or the peculiarities of labour, e.g., an inherited caste system.

The changeover from predominantly placer mining to large-scale lode production necessitated the investment of hitherto unheard of amounts of capital in mining. This turn of events, as it would happen, coincided with the generalisation of the joint stock company. Capital, in whatever size required, would henceforth be available for any investment, nay, would be seeking opportunities for investment. 'Joint-stock enterprise... has swept up all... available resources. Like a gigantic system of irrigation it first collects and then pours them through innumerable conduit pipes right over the face of the country, making capital accessible in every form at every point' (quoted in Trewhela, 1970, p. II/1). A form of capital adequate to the new demands of goldmining thus arose precisely when it was needed.

But the development of social capital was a two-edged sword. While huge amounts of capital becomes available, this capital, like all social capital, was flexible and fluid. It could therefore be applied as quickly and easily as it could be withdrawn. Goldmining required capital not only in enormous amounts, but also for very long periods of time. The fluidity of capital subjugates goldmining to the realisation demands of capital, so that whereas it at first appears as if capital is utilised by goldmining, it turns out that goldmining is utilised by capital. Goldmining becomes an application of capital like any other, and most other applications have shorter time horizons. Production is now determined not by the technical development of the mine, but by the need to pay dividends in order to hold onto capital. Speculation, especially prior to the imposition of stricter regulations on stock exchange activity, was the undoing of otherwise promising goldmining industries in Virginia, Colorado, and elsewhere in the United States.

> The gold interests of Virginia has suffered much from speculation. A party secures specimens, hurries to Northern cities, gets up a company of speculators, who develop the mine enough to sell to another company of speculators, and so the business has been conducted. In California and other new countries, men who go there to mine have to depend on their daily work for support, and are forced to work systematically and industriously, or fail in supplying themselves in food and clothing. If this had been the case in Virginia, instead of buying and selling mines on speculation, the gold interest would have been in a very different condition. In Virginia the deepest mining for gold has been not over 150 ft., while in California shafts have been opened 1,000 to 1,400 ft. ...We have gold mines in Virginia which will pay liberally if properly worked and managed. The many speculating schemes which have been inaugurated and failed, have made people adverse to believing in the gold belt of Virginia; but now that the gold yield of California is falling off, and capital there invested is not yielding so satisfactorily, capitalists are turning their attention to Virginia, and a fresh impetus has been given to this industry in our state (Pollard to Lock, 1881, quoted in Lock, op. cit., p. 183).[3]

Towards the end of the century share-holding for the sake of dividends became more representative. By then, though, and despite the destructive speculative activity of the 1860s and 1870s, mining was making steady headway in its transformation into a modern industry. Some metals, e.g., copper and iron, received a boost from their central role in the general process of industrialisation.

The industrial revolution in mining[4]

As discussed in Part III of the Introduction, it is inherent in the technical nature of mining that it admits mechanisation much more tardily than do other industries. When, by the middle of the nineteenth century, virtually all production had been set on an industrial footing, mining remained pre-industrial. Only in the final third of the nineteenth century did mining become mechanised. This means that whereas for most other industries the first and second industrial revolutions occurred more or less a century apart, for mining these two industrial revolutions coincided. The mining workforce, now, too, became an industrial workforce, but one which had skipped the manufacturing stage altogether.

> Mining reached its most dynamic innovation period towards the end of the nineteenth century. Prior to 1850, mining had not changed for hundreds of years. Beginning in about 1850 industry as a whole began to expand with the extension of railroads. The demand for metals and minerals grew enormously with increasing use of machinery and new inventions and innovations in industry as a whole. The mineral industries expanded at double digit annual compound growth rates. During the last fifty years of the nineteenth century the industry attracted talent and risk capital (as opposed to debt capital) (Douglas, H., 1987, pp. 268-9).

By the final third of the nineteenth century, that great invention which revolutionised all industry, the steam engine, was also in use in mining. Its underground application, though, was confined to pumping and hoisting, the engine itself remaining on the surface,[5] all other work underground was still being performed manually. Drilling was done almost entirely by hand. Cocopans were pushed by men or hauled by mules. In any case, these were small, being based on hand drilling and small-scale production. Hoisting was done in buckets, and (at larger shafts) on cages. Skips were used on incline shafts.

The more extensive and intensive use of steam power, together with improved engineering, the increasing substitution of planning and preliminary investigation for guesswork and trial-and-error procedures and a workforce consisting increasingly of adult males rather than children, resulted in improved productivity. These improvements were forced onto mining by the progressive exhaustion of ore bodies closer to the surface, necessitating deeper, more complicated and more expensive mining. But it was also facilitated by the greater amounts of social capital available.

Exploration was very primitive, consisting of panning, trenching, test-pitting and the digging of shafts and tunnels (these last two were often crooked). The dial compass and the dip needle were the first scientific instruments introduced. Their use was perfected in the 1890s. Drills were also used in exploration. Accurate mine maps were a rarity in the '70s. Mine mapping made a major advance with the introduction of uniform proportional scales[6] on the Witwatersrand.

From its first application in mining in 1865, explosives underwent a long, slow and hazardous process of development. By 1915, the ideal mining explosive had not yet been invented. Many of the changes in mining practice introduced in the 1870s, resulted from the introduction of dynamite.

In the 1870s, shafts were small, closely spaced, sunk directly on the vein and frequently crooked. This was because exploration had not yet sufficiently separated from exploitation to constitute an art in itself. Digging horizontally (drifting) was slow, meaning that shafts had a restricted operating radius of about 100 m. This necessitated shafts close to the ore and closely spaced. Thus has it been for a very long time. New technology and technique in horizontal development, especially the introduction of the jackhammer,[7] greatly increased the speed of this work. Ore could now be reached more easily by digging horizontally from an existing shaft than by sinking another shaft from the surface. This allowed shafts to be constructed further apart and their size and capacity to be increased. They could have more compartments each of larger size. Shaft linings changed from timber to steel and concrete.[8]

This achievement conceals the fact that shaft-sinking itself was one of the most difficult phases of mining operations to mechanise (Eaton, op. cit., p. 53). Prior to the introduction of the jackhammer to mining, shafts were sunk by hand drilling, which typically gave a sinking rate of 10 m per month in 'reasonably hard ground'. Jackhammers increased this to 130 m per month. Double-decker work platforms allowed a group of men on the lower deck to drill, while another group above built the lining. Improved fans (first introduced in 1835)[9] and pumps meant that the dust after a blast could be cleared much more rapidly and sinking resumed.

These developments meant that mining could proceed on a larger scale than had hitherto been possible. Certain new techniques, in themselves more costly than their older equivalents, proved economical in the context of large-scale production.[10] Mechanised systems began to appear underground in 1918 with the revolutionisation of tramming. Tramming is the operation whereby rock broken or blasted from the face is collected, cleared away and conveyed to the shaft for hoisting to the surface. Up to this time, tramming was an extremely labour-intensive part of the operations, accounting for much of the value of the eventual product. It must be kept in mind that the productivity of labour in drifting and stoping had been very greatly increased after the introduction of dynamite in 1865, and various ingenious ways of taking advantage of gravity underground.[11] By the teens of the twentieth century, tramming had already developed to where ore and rubble was shovelled or chuted[12] into cocopans running on tracks, these pushed by men or hauled by beasts (or, as in the coal mines of Victorian England, by children).[13] A scraper system was now introduced by which large boxes lying on their sides and tied in series to chains were dragged along the floor scraping up broken-off material. The scrapers were capable of being hoisted to a point were most convenient connection might be made with the shaft. This resulted in a one hundred percent increase in productivity. Shafts became more efficient in that the area they served could be anything from doubled to quadrupled, meaning an even greater distance between shafts. Repair work was much reduced in consequence of the more rapid extraction.

At this point, where the ore was transferred from horizontal conveyance through the galleries to vertical conveyance up the shaft, labour expenditure was particularly high. At the more sophisticated mines in the 1870s, cocopans were directly tipped into skips, while at the more basic mines this was done by hand. In either case, the cocopans were pushed up to, and back away from the skip (which was on the cage in the shaft) by hand.[14] Where hoisting was controlled with push-buttons, 'one man can load and hoist several thousand tons of ore per shift without overexertion' (ibid., p. 75).

Of course, this presupposes the use of electricity as a power source. Electricity began to radically alter the nature of mining when electric locomotives were applied to underground haulage from 1883, although the two forms of power, steam and electricity, were used side by side for decades. The internal combustion engine was first used in 1886. The old system of candles held in place on the brims of miners' hats by balls of clay, and which later gave way to various kinds of oil-burning lamps, was superseded by floodlighting in many parts of the mine, taking advantage of the presence of powerlines to feed the trolley locomotives. The use of electricity underground became generalised after 1910 (Douglas, H., 1987, p. 182).

Repair work itself was set onto a scientific footing. Up to about the turn of the century, e.g., drill steel was sharpened by hand, and 'every blacksmith had his own ideas about the proper shape of bits' (Eaton, op. cit., p. 65). The cutting angle and shape of bits, now scientifically determined, both reduced wear and increased cutting speed.[15] By as late as the early 1870s, the backwardness of pumping technology still caused mining operations to be shut down. The problem encountered at Lightning Creek [British Columbia] was not untypical. It is reported that, 'so much water is frequently met with that the pumps are mastered, rendering necessary a cessation of work till the driest part of the season, or the application of more powerful machinery.... In October of 1876, the quantity of water being raised amounted to about 13,870 gal. a minute' (Lock, op. cit., pp. 42-3).[16] The 3.5 kg/cm^2 managed by pneumatics when it was first introduced to mining in the 1870s, made it not yet practical for many applications. Over the next few decades, this pressure was trebled. Together with the improvements in transmission line design and manufacture, the concentration of great power on a small area became available ever deeper into the mine.[17]

Improved transportation and the spread of the metal-working and metallurgical engineering industries with large commercial machine shops, meant a reduction in in-house repair facilities. Standardised machines and parts provided by the metal industry further contributed to this development. Improvements were made both in steel and its treatment.

This overlap with industry in general manifested itself also in other important areas: power supply and labour. Steam, a universal source of power, lent greater freedom of site of production than did earlier sources, e.g., water. It also permitted the application of great force at a concentrated point. But this source of power still had limitations: fuel had to be brought to the site of generation; the storage potential of this power was limited (flywheels); and so was its transmission. Its application in mining was constrained by the steam engines having to be located on the surface, so that while it proved very successful in hoisting, it did not in drilling, tramming, haulage, ventilation, etc. Its application in pumping was only partially successful. But most significant of all, each user of steam power generated his own supply. Electricity was regionally generated and supplied, its transmission potential being infinite. At one level this meant that power could now be brought into the furthest recesses of any mine, but at another, it meant that mining became integrated with industry in general, as would be any other industry.

This integration into industry in general was not only in respect of power. Mining came to rely on the new and rapidly developing chemical industry for one of its most crucial means of production, dynamite. According to one source, 'Improvements in explosives were another innovation that pushed mining costs down; however, the research and development was undertaken by chemical companies, not mining companies' (Douglas, op. cit., p. 183). Mining was taking its place in the developing general symbiosis between industry in general as units of production, and industry in general as units of consumption.

> Important ...was the deployment of genuine industrial techniques. Ore was extracted with dynamite, shafts 500 m deep were sunk, and above all, in 1890 MacArthur and Forrest developed the cyanide process which replaced mercury and which made it possible to extract all the gold present in ore (Vilar, op. cit., p. 330).[18]

This integration went even further in that mine workers were becoming general industrial workers, capable of working in almost any industry. The technical requirements of mining (machinery, fuel, operating practices, etc.) had to move closer to those of industry in general. The workers were free wage-labourers, and hence the mines competed for the same labour with all other industries. Its labour requirements had to increasingly approximate those addressable by the average industrial worker.

Up until the last third of the nineteenth century, 'knowledge of mining and skill in its performance were handed down from father to son by the apprentice system, and the son learned to do things in the way that his father and grandfather had done' (Eaton, op. cit., p. 40). This constancy was reinforced by the extremely hazardous character of underground work. An ideology of anti-innovation and superstition pervaded mining circles.[19] This formed one of the stumbling blocks on the way to introducing industrial methods in mining. Having become part of a free wage-labour force, the average standard for wage-labour (by all criteria) would also become the standard for goldmining labour. Requiring a workforce of industrial standards of education and health, changerooms were, for the first time, built at mines to enable the workers to wash and change from their wet mining clothes into dry street clothes at the end of work. But the prevailing ideology of the the workforce holding that bodily filth was a sign of strength and virility, these facilities were shunned and, in some cases, burnt down (ibid., p. 80).

The vastly increased scales of mining operations required more men than the apprentice system could provide. The steam engine as a universal source of power necessitated new equipment and new working methods. For the first time, training institutions were founded through which, in addition, new ideas were disseminated.[20] The rate at which new ideas were introduced into mining once the ideological stranglehold of 'it is safer to do as has always been done' had been broken, may be likened to that of water dammed up behind a wall suddenly breaking through.

Even its capital requirements threw it into the same scramble as all other industries seeking capital. Mining capital therefore had to begin to approximate the attributes of the average capital in the market. Under the driving force of social capital, all capital and all labour strive towards universal applicability. 'By the end of the nineteenth century', according to one source, 'the structure of the mining industry was developing along lines similar to other industrial enterprises of the period' (Douglas, op. cit., p. 269). Mining, as an industry, was well on its way to losing the special character it had held onto for centuries. This last point becomes particularly strong in the light of the higher standard of refining which has become the accepted average.

Refining

Above we have described the process of ore reduction as distinct from gold extraction. A further differentiation is now introduced: that between extraction and refinement. Gold occurs in nature in association with a number of other metals, notably silver, copper and lead. The relative concentration of gold *vis-à-vis* other metals varies considerably from ore body to ore body. When gold is extracted from the non-metallic ore, other metals are usually extracted along with it. Some processes require that the gold first bind to an extraction agent to assist the process. In both cases the gold emerges in an impure state. The removal of associated metals and added chemical agents is the purification or refinement of the gold.

Fineness is expressed in parts gold per thousand parts or in carat. The former convention, which allows for more accurate assaying, is employed in official and serious circles. Generally the minimum gold content for an article still to be regarded as made of gold is one third (8 carat or 333.3 fine). Historically, 'pure' gold has been 916 fine (22 carat). This was the gold of the international monetary system in the nineteenth century. At 1,000 fine gold becomes brittle. Where the highest purity is strived for, therefore, the object is to add the minimum impurities that would prevent the onset of brittleness.[21]

Table 2
Gold finenesses and their uses

carat	fineness	expressed as...	comment
24	1,000		100% pure, uneconomical to achieve, brittle
	999.99	.99999	electronics
	999.9	.9999	purest monetary gold-20th century
	995		'good delivery' (min.) purity for London mkt
22	916		investm. jewellery, mon. gold-19th century
18	750		high quality jewellery
14	583.3		
10	417.7		
9	375		
8	333.3		minimum acceptable purity for jewellery

Fineness has its own value implications. To extract gold which naturally occurs at 875 and gold which naturally occurs at 975 might cost the same amount of labour-time per unit weight. One kilogram containing less gold (875) might therefore have the same amount of value as one kilogram containing more gold (975). Either one, or both, of two things can result: (i) The value of 975 gold is simply reduced to that of 875 gold. 975 will go out of circulation, be adulterated, and re-enter circulation as 875. The new 875 gold will now have a higher value than the old 875 gold, since labour will have gone into the adulteration process. In the first instance this additional labour is wasted, since the socially necessary labour for the extraction of 875 is already established. However, a portion of this additional labour is salvaged in that the baser metals used in adulterating 975 gold to 875, now circulate *as gold*. The weight will now be slightly higher than one kilogram, necessitating the removal of some material which would, of course, give a surfeit of 875 gold; (ii) On the other hand, should 975 gold be that which legally circulates as money, 875 gold will be refined to 975, This, too, would involve an additional labour input. The new 975 would therefore have a higher value than the old 975. Since the value of 975 is already determined, this additional labour, too, is wasted. This is, however, partially compensated for in that all the 875 gold in the one kilogram unit is now acceptable at 975 value whereas previously none of it was so accepted. The weight would now, of course, be slightly less than one kilogram, necessitating the addition of more 975 gold.

The value of gold, by the turn of the century, came to fluctuate in response to the following variables: (i) labour input; (ii) technical input; (iii) grade (iv) fineness (every time it becomes technically possible to improve the fineness, the value is increased). Improved fineness, and greater precision at attaining it, are achieved through complex chemical processes carried out in industrial plants on the surface.

The general tendency for the value of commodities to decline as an expression of the increase in the productivity of labour appears, in the case of extractive industry, as a *counter*-tendency. As the resource becomes increasingly depleted, so must the productivity of the labour producing it decline. Unless technology can be applied to counter this tendency, the intensity of labour has to increase. We already know from the nature of the money-commodity that the labour involved in its production is applied at its maximum intensity.

Although it might seem as though the value of the products of extractive industry is determined by different rules from these of other industry, this is but a perceptual trick resulting from the phenomenon being viewed from a different vantage point. The elementary movement is for labour to become increasingly efficient.[22] The unavoidable objective condition in extractive industry is that every application of labour immediately renders its successive application less efficient because the resource extracted becomes more thinly spread. This does not, of course, happen in an even and linear fashion. New discoveries of sources can and do dramatically reverse this trend. So, too, do major technological revolutions. Though a resource may be fairly evenly spread on a global scale, this does not mean that it is evenly spread at its actual site. A resource may occur more richly near the surface than it does at depth; one type of ore may yield it more readily than does another; etc.

The best of scientific predictive capabilities notwithstanding, labour in extractive industry only knows its efficiency *after* the production process has been completed. But then one must ask, 'what, exactly, does this labour do?' Let us consider the case of gold in particular, although the same principle would apply in any extractive industry. Strictly speaking, such labour cannot be said merely to be producing gold. Two distinct productive activities are taking place: reduction and extraction. Reduction involves everything from the first assault on the ore body, be it with hammers and chisels, dynamite, jackhammers or computer-controlled multiple-head drilling rigs, through bringing the material to the surface, to pulverisation to the point where no further mechanical action upon the ore will serve any purpose. Extraction involves a sequence (usually) of processes — mechanical, chemical and/or electro-mechanical — of segregating the gold from everything else. The richness with which the gold occurs in the ore, and the evenness of its distribution within it can only be accurately known between the end of the reduction process and the start of the extraction process. The point is that this richness (grade) is given by nature. Production, therefore, is more a business of *exposing* the material than of *bringing it into existence*. The question of the efficiency of labour has a different meaning depending on whether one is dealing with reduction or extraction.

Although overall efficiency can be expressed as x amount of labour-time to produce y amount of gold, to express it simply like this is deceptive. The two components of the production process must be considered separately, since the objective constraint that there is only so much gold in a given body of ore affects neither the efficiency in reduction, nor that in extraction in and of themselves. There is a difference between one thousand men reducing one hundred tonnes in ten weeks, one thousand men reducing ten tonnes in ten weeks, and one thousand men reducing a thousand tonnes in ten weeks. No matter how efficient or inefficient ore-reducing labour, there is never any gold at the end of the process, just pulverised ore. Efficiency thus reduces itself to either how rapidly a definite quantity of ore can be reduced to this state by a definite number of workers; or how much ore can be reduced to this state by a definite number of workers in a stipulated time; or how small a number of workers can reduce a definite amount of ore to this state in a definite time. Clearly, the application of ore-reducing technology has a positive effect on the productivity of ore-reducing labour, and so, on overall labour. The efficiency with which the ore is *reduced* responds proportionately (positively) to the general movement of the productivity of labour in the economy as a whole, since it does not produce gold, but reduces ore. But, the *overall volume* of ore processed in a given time may be increased not only by making the labour more efficient, but also by increasing the number of labourers at a given level of efficiency. This amounts to an expansion of the scale of production.

Extraction, on the other hand, begins from an altogether different premise. The process receives pulverised ore from the reduction process in such condition that the gold may now be extracted from it. Here, too, of course, can the productivity of labour be improved by the introduction of improved technology. But it only has so much gold as is contained in the ore handed over to it. Assuming the gold to be 100% extractable, improvement in productivity may be expressed either as a reduction in the number of labourers required to extract the gold in a definite time, or as a reduction in the time needed to extract the gold. But never as an increase in the amount of gold extracted by a definite number of workers in a definite time. Of course, it can also be expressed as a definite quantity of reduced ore processed in a stipulated time, but this would tell us nothing of the gold yield, which is the whole point of extraction. Paradoxically, independently of technological improvements, the total volume of reduced ore processed increases the total amount of gold extractable, but this tells us only about either the intensity of labour or about its efficiency.

The 1890s saw a complete revolution in refining, especially with the introduction of the MacArthur-Forrest cyanidation process in South Africa. By the turn of the century, the product of the milling process, a gold-bearing suspension, was put through mercury amalgamation, chlorination and cyanidation. Each of these was a distinct chemical process requiring a complete plant. When Douglas says that the MacArthur-Forrest process made Wits gold economic to mine, he confuses a technical problem with a value problem (Douglas, op. cit., p. 214). Without the MacArthur-Forrest process, the gold would simply have been irretrievable, regardless of how much money was thrown at it. The profitability or otherwise of the Witwatersrand goldfields was a value, rather than a technical question, which is the subject of section III. The point here is that by the turn of the century, actual mining, as traditionally understood, had been reduced to a fraction of the total mining productive activity. More on this in a moment.

Generalised wage-labour

One of the many landmarks in social evolution to which the nineteenth century may lay claim is the abolition of slavery. Although the coherent campaign known as the 'abolitionist movement' can be traced back to the eighteenth century, the process of abolition was uneven, riddled with counter-campaigns, often reversed and seldom altruistic. This notwithstanding, it is generally recognised in the literature that by 1850, slavery had been, for all intents and purposes, eliminated as an economic factor in the global economy, although it persisted as law in the United States until the Constitutional Amendment of 1865. Slavery continued in some peripheral areas well into the twentieth century. To venture into this vast and, apparently, under-explored question is beyond the purpose of this thesis. Interested readers are directed to the large body of literature.[23] We are interested in this question only in so far as it has a bearing on the revolution in the production of gold.

There is debate over the precise manner in which the rising industrial capitalism dovetails with the declining slavery and slave trade. Those who have sought a simple one-to-one relationship between the demand for a universal population of wage-labourers and state-sponsored abolitionism, at least in the context of the British Empire, have invariably found the situation far more complex than they had supposed. Either way, the end result was the universal demise of slave forms of labour, and its replacement by proletarian forms.

While this changeover was profound, it was by no means rapid. While western slavery was a major social force in the first half of the nineteenth century, it was nevertheless a form of labour existing in the service of capital. The slave trade and slavery were themselves a great boom to capitalism. The abolition of slavery was not so much a social revolution as a social adjustment. Neither were great classes wiped out, nor did new ones spring to life. A process which had been underway for more than one hundred years was merely being pushed to its logical conclusion.[24] This adjustment could not be immediate for all capitals. Plantation owners, e.g., howled about the certain collapse of the British Empire (or, at least, its sugar-growing sector), if they could not continue production on the basis of slavery.[25]

In order to ease the transition from slavery to wage-labour, intermediate forms of labour were introduced. The most important of these were 'apprenticeship' and indenture. The former was used especially to bind nominally free ex-slaves to their former masters purportedly in order to provide 'training' in the ways of wage-labour; while the latter involved essentially a long-term labour contract with the colonial state (breach of contract was a criminal, rather than a civil offence) and strict control over the value of the labourer's labour-power. These transitional forms were introduced as, where and for as long as required, hence apprenticeships in the Caribbean and the mass migration of indentured labourers both into and across the British Empire (see, e.g., Hobson, J., 1938, p. 250).[26] The transitional nature of indentured labour as a form lies not only in the fact of the freedom of the indentured labourer ranking somewhere between that of the slave and that of the wage-labourer as a consequence of which the value of his labour-power, too, lies between that of the slave and that of the free wage-labourer. Indentured labour is transitional also in that the indentured labourer, while bound to a particular employer for a given length of time, was nevertheless scientifically allocated to production areas around the globe according to a colonial central plan. 'They [the imperial nations—JH] induce an ever-growing stream of labour to flow between different parts of the subject portions of this Empire' (ibid.).[27]

The important point, though, is that henceforth, the labour of the world would become an increasingly integrated mass. The prominence of local attributes would gradually give way to average world attributes, though the process would be uneven and non-linear. This unevenness appears most starkly in the rapidity with which labour in different sectors begin to approximate their average, as against the sluggishness with which this levelling occurs between different countries. Thus does labour, through the process of universal proletarianisation, come to compliment the fluidity and flexibility of capital attained by social capital. No sector can any longer isolate its capital from capital in general, neither can it its labour. The same applies to gold production. This is, of course, the question of the influence of abstract labour on concrete labour.

This is a question which much exercises Ticktin's mind (op. cit., *passim*). For Ticktin, abstract labour works its way back into concrete labour in a different way. Capital seeks the most advantageous conditions for its realisation, and to this end continually shifts from investment to investment. For productive capital, this means from industry to industry. While capitals in direct competition with one another (such as in the same industry) and under the influence of that competition drive towards a homogenisation of their labour-power and its exercise, this occurs only indirectly across the entire productive landscape. Within an industry, therefore, the capacity of a unit of labour-power to yield surplus-value tends towards an average for the industry. There is a tendency for the statistical 'man-hour' to take on a fixed meaning for the industry.

Industrialisation allows for the drawing of all industries into such single universal determination. This occurs through the increasing standardisation of the factors of production. Each industry providing means of production to others is either subject to similar competitive pressures and hence, too, operating with labour-power of standardised attributes, or operating as a private or public sector monopoly in which case, of course, all users receive a product produced under the same conditions. Labour-power, similarly, tends towards an average across the economy. Industrialisation demands a free flow of labour across industries. The result is, 'the social homogenisation of labour. Its consequence is that a fluid, competitive and flexible workforce is created and maintained' (ibid., p. 56). This means that labour-power must have such attributes and be of such a standard as to be universally applicable. The need to measure social labour devolves away from money and, under the influence of industrial capital, back to labour itself. Labour-power then becomes directly its own unit of account across the economy. Or, as Ticktin puts it, 'abstract labour produces the abstract labourer' (personal communication, 2. vii. 93).

One may see, 'the social reduction of labour to a common form' (Ticktin, op. cit., p. 5), as doing for living labour what, more elementarily, it does for objectified labour. The qualitatively different concrete labours objectified in commodities are rendered commensurable by their social form of abstract labour. Abstract labour tries to render living labour similarly commensurable. Indeed, from the point of view of economy in general, commensurability of *living* labour suddenly springs to the fore as the original object of abstract labour, since it is living labour which an economy has to apportion to its various productive tasks. The continual correcting and over-correcting of social labour allocation which is the market is a necessary imperfection in the evolution of that allocation process. Industrial capital itself strives to correct this imperfection.

The value of gold, like the value of everything else, is always determined by the value of the labour-power expended on its production. But gold is a homogeneous commodity, meaning that all gold has the same use-value, regardless of whether a particular body of gold has been produced from placers or from lodes. All gold, regardless of the quantity of labour-power expended on its production, will only have the socially necessary component of that labour-power recognised as value. Most of the world's gold production, up to the second third of the nineteenth century, was from placer sources. Indeed, it is from placer sources that this material was first introduced to mankind.

The average socially necessary labour-time for gold production was, thus, set by placer production. But from the moment the role of money-commodity settled upon this commodity, its accumulation, and hence its production, became an end in itself. All gold resources technically exploitable were exploited. But from none of these sources could gold be retrieved at the value of the necessary average, save, perhaps, that of plunder from hoards.[28] Thus was the second most important source of gold, lode mines, always characterised by labour unable to reproduce itself. As an instance of production, it was the complete negation of the purpose of production: consumption.[29] Lode gold was simply not producable at the value of placer gold, hence the consistent working to death (while placer gold was not producable at a lower value since placer production was inconsistent with working to death).

> Out of the 6,500 and odd millions acquired by the European world down to ...1810, less than 500 millions were obtained through commerce; the remaining 6,000 millions were the fruits of conquest, plunder and slavery. [7.5% by equivalent exchange; 92.5% by unfree labour —FS]
>
> ...Between 1810 and the present time [1900] there have been produced from the mines, chiefly those of America, Russia, Australasia, British India and South Africa, about 13,000 million dollars, that is to say, in a single century twice as much as in the three centuries previously. Of this amount, about 4,500 millions were obtained by means of slave, serf, peon, or 'contract' labour; while the remainder was mainly the product of free labour, chiefly in North America and Australasia, a small proportion having been derived from commerce with Asia and Africa. The product of Asia, outside of British India (the Mysore mines) is a negligible quantity, because it has nearly all been consumed in Asia, and in addition thereto 4,000 millions of western metal. [$^2/_3$ by free labour; $^1/_3$ by unfree labour -FS]
>
> Taking the two periods together, the general results are as follows: from the Discovery of America to the present time the European world has acquired 19,500 and odd millions, of which

1,000 millions were obtained by conquest, 9,500 millions by slavery, and 9,000 millions chiefly by free mining labour (Del Mar, A., 1969, pp. 447-8).

One of the problems with the fixed gold price, is that it was fixed at a level far lower than the value of the labour-power needed to produce it under late-nineteenth century conditions. This is a consequence of the average social labour-time necessary for the production of a unit of gold having risen with the changeover from mainly placer to mainly lode mining on a world scale. For reasons internal to the workings of the international monetary system at the time, this change could not be allowed to feed through to the gold price. Henceforth, gold would have to be alienated below its value.

Producing below value is not a self-reproducing condition. Labouring below value, whatever the particular production engaged in and whatever the form of economy, must necessarily be premised upon unfree forms of labour. Such unfree forms were incompatible with the development of free wage-labour, the general movement of the time. Commoditised labour operates on the principle of exchange of equivalence. Around the world goldmining operations were either discontinued or not even started on account of the value of the labour-power being higher than that represented by the fixed price of the gold to be retrieved by that labour-power. If the price was not going to be adjusted to reflect the value of the material, its continued production depended upon one of two developments: (i) either the technological basis of gold production was revolutionised; or (ii) a source of labour-power is found the value of which would be lower than that represented by the price of the gold it is to produce.

While that portion of the precious metals (about one half of the existing stock), which was obtained through conquest and slavery, practically cost nothing, the portion obtained through free labour cost more than it is worth: a fact long familiar to mining men and now admitted by numerous publicists. Such is the penalty which nations must pay for indulgence in violence and crime: their fruits lower the value of the products of free labour. Not until the precious metals have entirely ceased to be acquired through conquest and slavery can free mining labour hope to obtain an equitable reward in the value of its product (ibid., p. 237n23).

With a revolution in the value of labour-power, a complete readjustment of all value relations takes place.

New value-conditions

To distinguish placer surface work from placer deep mining is useful in that placer gold production, in the latter case, assumes all of the technical constraints associated with mining in general. To distinguish placer mining in general from lode mining and conglomerate mining, is to distinguish between (i) relatively very high yield production vs relatively very low yield production; (ii) the individual miner as unit of production vs a company as unit of production; (iii) labour-power being the largest factor of production vs equipment being the largest factor of production; (iv) production possible on the basis of labour alone and production impossible without capital. But there is a further critical distinction to be made, and that is between the emergence of a finished product (native gold) after the *mining* process, and the need for further *non-mining* production processes to secure the finished product.

The value of lode gold has not only been higher than placer gold on account of its need for digging and tunnelling before the gold is reached, but also on account of its need for the extraction of the gold from the ore, after the ore had been reached (the ore, of course, also has to be brought to the surface). The standard process for centuries has been that of amalgamation. This involves exposing the ore to mercury, to which the gold particles adhere. To ensure adequate contact, the ore had first to be pulverised. Much gold was lost in this way, and, since the value of gold had always been set by surface placers, any gold production involving even a small amount of amalgamation necessarily never repaid the labour-power that went into the total production process. At some placer sites approaching exhaustion, amalgamation was even applied to the ore because the remaining particles were so minute and light that they could not be segregated by the same crude physical processes.[30] Amalgamation is a refining process. The depletion of the placers means that, on a world scale, this process became an integral part of the production process which delivered the bulk of the world's gold. Native gold produced by digging alone has shrunk to an insignificant portion of world output. For the remainder of output, 'the concrete labour of gold-digging' produced no gold at all.

This means that for all time up to the last third of the nineteenth century, one may speak of a placer period, followed by an amalgamation period, in gold-production.[31] Finley identifies a further, 'cyanide and smelting period' starting in 1890. This further identification, though important in the development of extraction technique, is not important in political economy, as it continues the same development as that brought about by the rise to prominence of amalgamation, rather than introducing a further production revolution.[32]

The relative importance of post-mining production operations in goldmining by the late nineteenth century may be gleaned from the fact that of the total Witwatersrand gold production in 1898, 47% of the gold emerged after mercury amalgamation and 53% from a further series of chemical extraction processes, viz. chlorination and cyanidation (WCM, 1897, Tbl. 'Monthly Returns and Averages for the Year 1897', no page number). Native gold was coaxed from the ore *in plants on the surface*, i.e., under technical conditions comparable to those in any industrial plant. The value-relation is even further complicated by the fact that the workers down the mine, who, at the end of their digging, produced only broken rocks and dust, made up 70% of the workforce, while those in the chemical plants actually winning the gold made up only 8.67% of that workforce (ibid., Tbl. 'Labour Returns', no page number).

The importance of this point is that 'the concrete labour of gold-digging' has, in theory, become superfluous to gold production, since it is, as concrete labour, identical to what goes on in a quarry. The increasing miniaturisation of industrial equipment, infantile as this tendency may have been at the end of the nineteenth century, meant that, inexorably, mechanisation would bore its way ever deeper into the mines, eventually bringing high-efficiency machinery right up to the face. The establishment of complete processing plants underground, with only the finished product sent up to the surface, is therefore not only a theoretical, but also a practical possibility, albeit distant. The value of gold, like the value of everything else produced under industrial conditions, becomes a very complicated matter.

Only after the completion of eighteen months of empirical research to substantiate this thesis, did we discover, quite by chance upon a rereading of the relevant chapters in Volume I of Marx's *Capital*, that these developments are, indeed, mentioned in this work — not by its author, but by the editor of its fourth German edition, Engels. In the latter edition, published in 1890 (seven years after Marx's death), Engels adds a lengthy passage to

Chapter Three in which he describes precisely that series of developments in gold production which we have submitted to examination in this study. It is necessary to quote the passage in full:

> We find ourselves once more on a period of serious change in the relative values of gold and silver. About 25 years ago the ratio expressing the relative value of silver and gold* was $15^{1}/_{2}:1$; now it is approximately 22:1, and silver is still constantly falling as against gold. This is essentially the result of a revolution in the mode of production of both metals. Formerly gold was obtained almost exclusively by washing it out from gold-bearing alluvial deposits, products of the weathering of auriferous rocks. Now this method has become inadequate and has been forced into the background by the processing of the quartz lodes themselves, a way of extraction which formerly was only of secondary importance, although well known to the ancients. Moreover, not only were new huge silver deposits discovered in North America, ...but these and the Mexican silver mines were really opened up by the laying of railways, which made possible the shipment of modern machinery and fuel and in consequence the mining of silver on a very large scale at a low cost. However, there is a great difference in the way the two metals occur in the quartz lodes. The gold is mostly native, but disseminated throughout the quartz in minute quantities. The whole mass of the vein must therefor be crushed and the gold either be washed out or extracted by means of mercury. Often 1,000,000 grammes of quartz barely yield 1-3 and very seldom 30-60 grammes of gold. Silver is seldom found native; however, it occurs in special quartz that is separated from the lode with comparative ease and contains mostly 40-90% silver; or it is contained, in smaller quantities, in silver, lead and other ores which in themselves are worthwhile working. From this alone it is apparent that the labour expended on the production of gold is rather increasing while that expended on silver production as decidedly decreased, which quite naturally explains the drop in the value of the latter. This fall in value would express itself in a still greater fall in price if the price of silver were not pegged even to-day by artificial means. But America's rich silver deposits have so far barely been tapped, and thus the prospects are that the value of this metal will keep on dropping for rather a long time to come. A still greater contributing factor here is the relative decrease

in the requirement of silver for articles of general use and for luxuries, that is[,] its replacement by plated goods, aluminium, etc. One may thus gauge the utopianism of the bimetallist idea that compulsory international quotation will raise silver again to the old value ratio of $15\frac{1}{2}:1$.** It is more likely that silver will forfeit its money function more and more in the markets of the world.[33]

This is the earliest mention of these developments in *Capital*, which confirms us in our view that Marx was unaware of them. Marx over-concentrates on the *relative* values of gold and silver in terms of each other, at the expense of the actual value of each in terms of its own necessary labour-time (see Marx, 1973, pp. 173-85 and 1977b, pp. 75-6, 156-7). His emphasis is on the circulation of the precious metals, rather than on their production. Their production is not entirely ignored, but such examination is merely ancillary to a discussion of the *Reproduction and Circulation of Social Capital* in Volume II of *Capital* (Marx, 1977a, pp. 474-8). While an analysis of silver does not form part of this study,[34] it has to be said that a concentration on the value ratios between the two metals must invariably relegate the question of their own intrinsic values to the background. It is most unlikely that an astute scholar like Marx should have fallen foul of such an oversight.[35] It was rather the case, as we suggested elsewhere, that his object was to study capital, and not the precious metals. He was therefore only interested in the precious metals in so far as they elucidated capital. Indeed, he explains that, 'we take it that the gold mines are in a country with capitalist production whose annual reproduction we are here analysing' (ibid., p. 474).[36]

Having thus set the parameters of his investigation, the production of gold and the reproduction of the gold producers only comes under scrutiny to the extent that these form 'direct elements of annual reproduction' in a gold-producing, fully-industrialised, capitalist country. This means that, both historically and geographically, gold production outside of the nineteenth century United States and, to an extent, Canada and Australia (which were colonies), have to be excluded. This is a restriction consciously elected by Marx, who concludes that, 'consequently, gold too is to be treated here as a direct element of annual reproduction and not as a commodity element imported from abroad by means of exchange' (ibid., p. 474).

The annual reproduction of the capitalist economy as a whole is not the subject of our thesis, but Marx's point of departure appears to stand in direct contradiction to his treatment of gold in *A Contribution to the Critique of Political Economy*. In Volume II of *Capital* he says that, 'the production of gold, like that of metals generally, belongs in class I, the category which embraces the production of means of production' (ibid.). In the *Contribution...*, though, gold is expressly excluded from means of production:

> Metals in general owe their great importance in the direct process of production to their use as instruments of production. Gold and silver...cannot be utilised in this way because ...they are very soft and, therefore, to a large extent lack the quality on which the use-value of metals in general depends. ...The precious metals are useless in the direct process of production (Marx, 1977b, p. 154).[37]

As Engels expected, the demand for silver as Use-value II, i.e., limitless demand, only persisted in those societies which still continued to employ silver as money, e.g., China.[38] Engels could not, of course, rework the affected sections of *Capital* on the basis of this new information, but he could have said something about its importance and that such rewriting might be necessary. That he did not mention this suggests to us that he did not fully appreciate the significance of his own insertion, apparently believing that it required only that Marx's facts be brought up to date.

Despite having the benefit of this very important information, Engels restricts his theoretical exploitation of it to the theoretical concerns of Marx: the silver:gold value ratio. Engels talks of the distinction between placers and lodes, the historical predominance of placers over lodes; the reversal of this relation in the late-nineteenth century, the arrival of machine production in mining, the decline of the value of silver simultaneous to, but independently of, the rise in the value of gold and the expulsion of silver from the monetary role. None of these is addressed by Marx. The form of ancient mining described by Diodorus Siculus, *and which serves to underpin part of Marx's theory*, is described by Engels as having been, 'only of secondary importance'. At the very least this should have alerted Engels to the possibility that Marx's conception of the money-commodity in general, and the measure of value in particular, might benefit from a reassessment. Instead, he directs his energies outwards, towards the bimetallist advocates, whom he criticises for seeking to maintain a fixed silver:gold value-ratio.

Del Mar, who is rather unkind to those whom he sees as blindly peddling the 'formula' of cost of production, here, and throughout his book, finds it impossible not to resort to it himself. At one point, e.g., he explains that:

> The present cost per foot of driving a gallery in the Wynaad mines of Mysore — this means blasting out and removing to the surface $7 \times 4 = 28$ cubic feet of rock — with 'native and Eurasian labour', is only 4s. or \$1. It is against this system of peonage that the free labourers of California, Australia, and British Columbia have to contend in producing gold at £4 4s. $11^{1}/_{2}$d or \$20.67 per ounce fine for the London, Philadelphia, and San Francisco Mints (Del Mar, A., 1969, p. 430).

He clearly feels very passionately about the value question, but he does not recognise it as such. Almost never, then, is gold *theoretically* examined in terms of the amount of labour-time necessary to produce it or in terms of its role as money-commodity. Detailed *empirical* studies have been done on this question (Lock, Del Mar, Jacob). Hirson, in a slight study, does correctly identify the parameters of the question. Seeking to explain the basis of the low wages of South African black mine workers at this early time, he suggests that gold produced by them came 'onto the market' at a, 'price set ...far below its value', as a result of which, 'every means was employed to ensure that workers were paid the lowest possible wages' (Hirson, B., 1993, p. 56). This is to view the problem from the wrong end, for Hirson suggests *competition*, a relation between capital and capital, as the appropriate category for examining a relation between money and capital. Given that this is but a slight study, and that the focus of his attention was elsewhere, it is perhaps not fair to point to the many theoretical (and factual) problems wrapped up in Hirson's conception. For almost all who write on early South African goldmining, the 'fixed gold price' is the first cause uncaused of South African goldmining economics.

Notes

1   'It was in this period that mechanisation first became characteristic of industry in general', quoted in Trewhela, P., 1970, p. 2.

2   '[One]... decisive consequence... [which] followed from the amassing of vast sums in the production of capital goods[,]... was an abrupt reduction in costs of production. In the 1880s the price of iron fell by over 60 percent or even more. The price of coal fell by over 40 percent. ...The amount of labour required for the production of a ton of rails had fallen by half since the middle of the century' (ibid., p. 2).

3   And further, from same letter, 'Many gold-mines have been worked in Virginia from time to time. Want of experience and of proper machinery, want of sufficient depth, speculative investments with no expectation of working them, except enough to get up a company and sell out, with a repetition of this by another company or individuals, have caused failures and suspensions of many of the mines' (ibid., p. 184). For the fate of the Colorado goldmining industry at the hands of speculators (see King, J., 1977).

4   The discussion here draws heavily on Eaton, L., 1948 and the Encyclopædia Britannica, Vol. XII, pp. 245-56. The concern is with underground lode mining as distinct from deep placer mining (see Introduction, Part III).

5   In open-cast mining, steam power had, of course, a wider application than in underground mining. From 1877, steam shovels running on tracks did the actual digging, bulldozers working in the parts which could not be reached by the shovels. Prior to the introduction of steam, all this work was done by hand. Churn drills were used in most open-cast mines, their application here following that for prospecting for deep placer deposits. Although draglines were used in loading and shifting placer ores, they were seldom used in open-cast mining.

6   1:100; 1:500; 1:1000, etc., as opposed to the then prevalent system in the English-speaking world of 1 foot to 1 inch; 40 feet to 1 inch, etc.

7   First introduced in the United States in 1897 (Douglas, H., 1987, p. 128). Although mechanical drills were first introduced in mining 1871 (piston drills driven by compressed air), neither these, nor their successors over the next 25 years, were really practical for mining. The jackhammer was followed by a hand-held self-rotating drill the material of which took advantage of the latest developments in steel

metallurgy. Heavier models of these were used for sinking shafts.

8   So rapid was the development in shaft design and construction, that by the mid-twentieth century, shafts were being sunk 3 km apart, giving each shaft an operating radius of 1,500 m.

9   By the late nineteenth century, ventilation was still by natural means, although assisted by fans and shut-off doors.

10 'The greatest improvement in development has been in large ore bodies in preparation for sublevel stoping and undercut caving. In these systems of mining the cost of development may be greater than that of stoping, but the overall cost is lower than in the older simpler systems, and production per unit area and per man employed is greater. Such systems would not be possible if we had to depend on old methods of drifting and raising' (op. cit., p. 55).

11 Instead of approaching the ore from above, as has traditionally been done, the mine is developed to the underside of the ore. Stoping (removing the ore body) therefore proceeds upwards and the broken material falls into position ready for removal. This obviates the need for a great deal of shovelling and hoisting work. There are many variations of the same principle.

12 Up to this time, though, most loading still continued to be done by hand. 'Loading by hand is the greatest drudgery in mining', says Eaton (p. 66). Power loaders were first introduced at about the same time as scrapers.

13 Mechanical haulage in the form of the continuous rope system already started appearing in mines early in the nineteenth century. Electric mine locomotives were introduced in about 1879.

14 By the mid-twentieth century, this procedure had been mechanised to the extent that a skip was being filled in three to four seconds.

15 From the mid-nineteenth century, metal-working, an art almost as old as production itself, gave way to metallurgy. This had many dramatic effects on mining. The quality of steel made phenomenal improvements in the latter half of the nineteenth century. Mining was able to move from soft iron and cast-steel ropes in the early 1870s, to steel of much higher tensile strength within a few decades. Ropes of

stronger steel meant that shafts could go much deeper as ropes could be much longer before they snapped under their own weight (before the introduction of intermediate underground hoisting stations). Hoisting engines therefore increased greatly in size and power, though these remained steam-powered well into the twentieth century.

16   Fifty years earlier, even this would have been impossible.
17   Improvements in steam transmission line design and manufacture (especially improvements in insulation) enabled the then rapidly developing steam pumps to lift water to a height of 900 metres in a single action. Electric pumps superseded these, starting with lifts of 1,100 m per action. These developments in drainage put previously inaccessible bodies of ore within reach of exploitation.
18   It is not quite true that cyanide replaced mercury. It rather came to complement it. Certainly by the outbreak of the South African War, both processes were being run in tandem. The substantial point, though, remains valid.
19   In English coal-mining communities it was considered bad luck for a woman to go down a mine.
20   This does not mean that the art of mining did not receive scholarly attention before. Georgius Agricola's *De re Metallica*, is not only famous for the detail of its content and the beauty of its manuscript, but also for its being out of character for many centuries both before and after its publication in 1556.
21   In the region of 99 and above, refining costs rise exponentially. There are enormous technical difficulties involved in purifying gold from .996 to .9999 fine. Both South Africa and the former Soviet Union went to extraordinary expense to purify their gold beyond .996 fine.
22   'Economy of time, to this all economy ultimately reduces itself', Marx, 1973, pp. 173.
23   The discussion here draws on Davis, D., 1984; Eltis, D. and Walvin, J. (eds), 1981; Mathieson, W., 1967; Asiegbu, J., 1969, who suggests that, 'the West African slave trade lingered on for many decades after 1841', p. 119; Williams, E., 1944, who is by far the most interesting since he examines the various diverse ways in which different capitals benefited from the maintenance of slavery or stood to benefit from its abolition;

Hobson, J., 1938; Encyclopædia Britannica, Vol. XVI, pp. 853-866.

24 Social changes appear to have been more profound in the slave supplying areas that in the receiving areas.

25 Those who sought to preserve slavery found that they did not have a monopoly on concern for the well-being of Empire. As one source reports,

> The West Indian [sugar] monopoly was not only unsound in theory, it was unprofitable in practice. ...England was paying for its sugar five millions more a year than the Continent. Three and a half million pounds of British exports to the West Indies in 1838 ...purchased less than half as much sugar and coffee as they would have purchased if carried to Cuba and Brazil. ...The capitalists, eager to lower wages, [argued that] ...monopoly was unsound, costly to all, and had destroyed the great colonial empires of the past (Williams, E., op. cit., pp. 138-9).

26 'One capitalist device for keeping wages low at least for a time is to bind immigrants to contracts before they leave the old economy. The Indian indenture system, for example, rests on such an arrangement' (Bonacich, E., 1972, p. 550).

27 It would be interesting to know whether this was the first *consciously planned* global allocation of labour.

28 The history of the value of gold is partially considered in Jastram, R., 1977, 1981, Lock, op. cit., Del Mar, op. cit., Jacob, W., 1968a, 1968b and Vilar, P., 1976. According to Jastram, R., 1981), the 'treasure' brought from the Americas consisted largely of silver. Silver circulated alongside gold as money and the stability of the value-ratio between then was of importance to economists at the time.

29 For an elaboration of the Production-Consumption duality, see Marx, 1973, pp. 90-94.

30 This was the case, e.g., in California, according to Vilar, op. cit., p. 326.

31 A similar periodisation is offered in Findlay, J., 1910, pp. 324-5.

32 Trewhela, P., 1986b, ascribes rather more significance to the introduction of the MacArthur-Forrest cyanidation process than we are persuaded of (see p. 45).

* The original reads, 'gold and silver', which is the wrong way round.
** The original reads, '1:15$^{1}/_{2}$', which is the wrong way round.
33 Passage added by Engels to the fourth German edition of Marx, 1983, p. 142.
34 For a discussion of the monetary role of silver in nineteenth century Europe, see Clapham, J., 1961, Ch. 13; for the United States in the same period, see US Govt, 1870 and Friedman, M. and Schwartz, A., 1963, Ch. 3; and for a general study, see Jastram, R., 1981.
35 Even a scholar subscribing to the misconceived quantity theory of money was able to ask,

> How could it happen that silver, depreciated by excessive production, should not fall in value in the very country where it is produced? ...For the simple reason, gentlemen, that silver has nothing to do with the result. It is the other train which is moving in the opposite direction. It is gold, your money of circulation in Europe, which you have made scarce, and which has therefore risen in value. It is in Europe alone that you have a crisis, due to a fall in prices (HMSO, 1893, p. 33).

36 Trewhela strongly argues that Marx was very late in his life, 'very specifically studying *the relation of gold production* to the total circulation process', pointing out Marx's use of Soetbeer, A., 1879 and a British Parliamentary report dealing with gold bullion and issued in May 1879 (Trewhela, P., 1986b, p. 27, emph. orig.).
37 This could make for a very interesting inquiry, especially since Engels could not find the study promised by Marx, and in which he would almost certainly have explained his reasoning. 'Supposing the annual production of gold is equal to 30. ...Let this value be divisible into $20c+5v+5s$; $20c$ is to be exchanged for other elements of Ic and this is to be studied later...' (ibid., pp. 474-5), at which point Engels adds the following footnote: 'The study of the exchange of newly produced gold within the constant capital of department I is not contained in the manuscript' (ibid., p. 477).
38 'Brazil, China, Spain and Turkey never made it [onto the international gold standard]' (Kemmerer, D., 1975, p. 110).

Part Three

THE VALUE OF LABOUR-POWER IN EARLY SOUTH AFRICAN GOLD PRODUCTION

Introduction

While Part I explored the nature of gold in its three aspects as use-value, money and capital, and Part II examined the detailed empirical developments which allowed the processes described in Part I to emerge, Part III will look at all of these as a complete process in a particular situation. The insights gained in Parts I and II will be tested in the context of early South African goldmining economic history. The primary purpose is to show that the peculiarity of South African goldmining flows from the processes and developments discussed in the previous two Parts. The secondary purpose is to demonstrate the explanatory potential of the category of value of labour-power, and thereby propose it to South African scholarship. This Part offers a political economy of South African gold-producing labour-power structured around the category of value of labour-power. It also lays the basis for a retheorisation of *apartheid*.

6 Competition in early Witwatersrand goldmining

The extent of co-operation and/or collusion in the South African goldmining industry invariably draws comment in the literature. The motivation for this co-operation/collusion is said to come from the need to control the cost of labour, while the opportunity for it is provided by the fact of all producers always selling the product at the same fixed price. One view holds that the collusion in goldmining reflects a particular stage of capitalism — that of 'monopoly capitalism'. According to this view, since both diamond-digging and goldmining arose in this monopoly phase, they will demonstrate the same developments towards monopoly. This view is most consistently represented by Duncan Innes. Another view holds that gold and diamonds have a 'different relationship between price and output', in the words of Frederick Johnstone. Either way, collusion is seen to find expression in the Group System of organisation by which the industry is characterised.

One inadequacy of the literature is that the presence of this collusion/co-operation is merely described empirically, rather than seen as part of a relation. That collusion/co-operation denotes the absence of competition is only partially brought out. An analysis of goldmining competition as such is not made. Capital being inseparable from competition, a question arises as to the *form* that competition assumes in this highly co-operative/collusive industry. Given the central role of competition in the apportioning of the total surplus-value between the different competitive capitals, how is this apportioning carried out for the goldmining industry — an industry which internally lacks competition. The form of competition in the South African goldmining industry has not, either, therefore, been taken up for study.

The object here is to attempt to uncover the particular form that competition has come to take in the South African goldmining industry, and to explain the manner in which the individual goldmining firms receive their portion of the total surplus-value. For this we shall be looking especially at two views of the nature of South African goldmining in its formative years. The first, that of Duncan Innes, is selected because it expresses a widely held conception of the nature of modern capitalism as 'monopoly capitalism', which is taken to be directly reflected in South African goldmining. The second is the view of Frederick Johnstone, selected because it is so often drawn on in the literature.

Duncan Innes' book *Anglo American and the Rise of Modern South Africa*, which offers an exposition of the monopoly capitalism thesis, also, devotes some attention to the 'conditions of competition' (Innes, D., 1984. See esp. pp. 45-74 for goldmining, and pp. 39-44 for diamond-digging). Innes' material is well researched and his position seems to be better argued than that of many. But he does not escape the inadequacies mentioned above. Innes' book, therefore, appears to be a useful starting point into the discussion.

The monopoly thesis rests on the theoretical conception that capitalism, by the late nineteenth century, had reached its 'monopoly phase'. New industries starting up in this phase, it is argued, are by their nature 'monopolistic'. Goldmining on the Witwatersrand, having started in 1886, does not escape this. The empirical side of the monopoly thesis rests on the fact that early goldmining capital was, to a very great extent, based on capital operative on the diamond fields. Diamond digging was, by 1888, already under monopolistic control. Innes sees monopoly as having been transferred directly from Kimberley to the Rand in the persons of the diamond magnates, who expanded onto the Reef (ibid., p. 47). The difference being that while monopoly *evolved* at Kimberley, it was present on the gold fields right 'from the very beginning'. At the same time, at the 'very beginning' of the Rand, Innes says, 'when the gold fields were first proclaimed in 1886 there were some 3,000 people active in the area. ...In most cases mining was carried on by a lone digger employing two or three Africans to help crush and wash the gold-bearing rock' (ibid., pp. 45-6).

It is true that large capitals which have cut their teeth on the diamond diggings moved onto the Rand at an early date, but it is incorrect to say, as we shall show, that they constituted a monopoly. Innes finds empirical evidence for monopoly in the existence of the Chamber of Mines and in the so-called Group System according to which the mines came to be organised.

The first problem with Innes is that he sees capital in diamond digging as forming a seamless continuum with capital in goldmining. We will show that these capitals are different by virtue of the different ways in which they interact with the market. The diamond market is a limited one while that for gold is unlimited. Innes himself acknowledges this. When describing an 80% increase in diamond sales between 1883 and 1887, Innes asks, 'how much longer could the market absorb the increase?' (ibid., p. 34). Later, when he describes the increase in gold production from 254kg in 1886 to 10,915kg in 1889 (an increase of 4,200%), the question of market saturation does not arise (ibid., p. 47).

This difference between diamond-digging capital and goldmining capital, in turn, can be shown to express the different social roles of their respective products. Diamonds, like all other commodities, stand as private use-values opposite gold as universal use-value. Diamonds are a particular form of wealth, like tomatoes and, indeed, gold. But gold is also the general form of wealth. Diamonds, etc. are needed for particular consumptive purposes, while gold, in addition, is needed by exchange economy itself as one of the objective conditions of its existence. It is the material incarnation of exchange value as such, the universal commodity.

It was on account of its social role as universal commodity that there exists an unlimited market for gold.[1] While Johnstone offers no explanation for the unlimited market for gold, Innes ascribes it to Britain's role as, 'the world's leading economic and political power' in the nineteenth century. Taking advantage of its position, Britain, 'bought up all the gold which was offered for sale on the international market', and was able to do so 'at a fixed price' (Innes, ibid., p. 48). It is true that London was the destination of virtually all Empire gold. But a question must then arise of what the other economic and political powers did to acquire the gold they, too, needed. If gold was the medium of international trade, and Britain was the 'leading economic ...power', others would buy more goods from Britain than Britain would buy goods from them. What purpose does the medium of exchange serve, if it is all concentrated into the hands of the seller of goods. Did Britain then resell the gold in order to then be able to sell its goods? Innes' conception derives from his failure, as it is the failure of almost all scholars, to grasp the concept of *universal* commodity, and to distinguish it from *particular* commodity.

As there was an unlimited market for gold and hence no need for a struggle over market share, there was no need for monopoly. Inconsistently, however, Innes equates an unlimited market with a glutted one. For him, 'conditions of competition similar to those which had prevailed in Kimberley in the late 1860s also characterised the earliest period of Witwatersrand gold production' (ibid., pp. 46-7). This is flatly contradicted one page later:

> The specific nature of the international market for gold had contradictory implications for the future of the development of the industry. On the one hand, conditions of an unlimited market at a fixed price meant that the gold industry would not be hampered as others, and especially the diamond industry, had been by market fluctuations and restrictions: the market for gold was unlimited and therefore the threat of over-production did not exist, while the existence of a fixed price meant that capitalists could plan production on the basis of assured returns on the sale of the commodity (ibid., pp. 48-9).

The tendency towards monopoly is capital's way of coping with its own 'inner necessity' — competition — and it is a way which, by undermining competition, undermines capital itself. But it needs only strive towards eliminating its competitors from the market once it becomes uncertain of its share of that market. This was the case with diamonds, but never with gold. Eventually, all diamond production did end up in the hands of one single producer, the De Beers Consolidated Mines Ltd. There, therefore, really was a monopoly in diamond mining (ibid., pp. 35-42).[2]

Competition and monopoly

The term 'monopoly', which describes a condition of the presence in the market of a single seller, is antithetical to competition — the presence of one indicates the absence of the other.[3] Monopoly, for Innes, appears to mean, 'very large and powerful companies operating internationally'. When referring to collusion amongst the world's largest diamond dealers, e.g., Innes talks of how, 'the monopoly Rothschild group... secured the support of other major dealers' (ibid., p. 36). The presence of such companies on a world scale also reflects a particular stage in the development of capitalism, so-called 'monopoly capitalism'. Any number of 'monopoly capitalists' can therefore compete with one another under conditions of 'monopoly forms of capitalist relations'. The opposite condition of which is one of 'petty-bourgeois relations of competition'. Innes has not abstractly explored either the relationship between monopoly and competition, or, for that matter, the relationship between capital and competition.

While competition has historically emerged out of monopoly — the monopoly of guild production — as a *development* over the earlier restricted forms, its return to monopoly in the present period describes, once again, a development over the wasteful *post factum* allocation of resources through the market. There was no scientific basis, from the point of view of production itself, for the guild monopolies. Production was even more atomised than it is under free competition, which at least strives to integrate production, albeit in a contradictory fashion.

The drive of capital towards monopoly stems not only from the need to dispose of competitors in order to gain control over price, but from the need to *rationalise its own production*.[4] This is best achieved where a capital controls all production for, and trade in, a particular market, i.e., under conditions of monopoly. In a sense, it can be said that society as a whole already exercises control over the rational allocation of resources in all production through the existence of social capital. Monopoly is, therefore, both a natural, and a *progressive* development out of competition. Having first appeared as guild monopoly, it has now gone through the full social development process of competition to re-emerge in a higher form, scientific monopoly.[5]

Monopolies certainly do exist in the world market and they do get their share of the total surplus-value. Innes does not explain how they could do this in the absence of competition in the market. Neither the question of surplus-value, nor that of monopoly rent are considered. Innes is able to reconcile competition and monopoly empirically by conceptually fusing them with oligopolistic rivalry. Hence he speaks of, 'organized collusion among capitalists in the industry' (ibid., p. 52), or:

> This did not of course mean that competition disappeared from the [gold] fields: on the contrary, it continued on a larger and often more vicious scale as monopoly capitalists sought to out-manoeuvre each other to gain possession of the richest areas on the fields (ibid., p. 47).

Innes might here legitimately have talked of oligopoly, in which case much becomes explicable. A small number of very large mining Groups (the 'Group' phenomenon is examined below) certainly competed very viciously on the fields, but not over disposal of the product, as was the case before monopolisation at Kimberley.

Competition persisted at other points in the circuit of capital, though. These were in the areas of securing finance, and securing factors of production (means of production and labour-power).

The fixed price and the unlimited market

The oft-cited work of Frederick Johnstone, *Race, Class and Gold*, unlike that of Duncan Innes, finds the basis for its thesis in the economics of gold itself — to whit, in its fixed price (Johnstone, F., 1976). Although Johnstone discusses 'the imperatives of goldmining on the Rand' in terms of three specific factors — these comprising, 'the low average grade of gold, the internationally fixed price of gold, and the high level of development and overhead costs' — two of these, the grade of ore and the development costs, would find their equivalent in any other industry. The differentiating factor must, therefore, be that which is unique to the goldmining industry: the fixed price. Johnstone himself clarifies it thus:

> The differentiating issue for the goldmining industry is not the relationship between the price of its product and the general level of costs with which it has to contend; in this respect, it differs in no way from other industries. What does differentiate it is that for long periods of time the price of its product has remained constant, though the cost level may have gone up (quoted in ibid., p. 18).

For Johnstone, 'the fixed and stationary price of gold had two significant consequences for the industry'. From the fixed price, Johnstone argues, follows: (i) 'that the mining companies were unable to transfer increases in production costs to consumers in the form of price increases'; and (ii) 'that the goldmining industry was not subject to crises of over-production' (ibid., p. 18). The first of these consequences is quite straightforward and requires no further clarification. But what does need explanation is how, exactly, a fixed price leads to a limitless market. Johnstone offers none for, indeed, there is none: the second is a *non sequitur*. As has already been shown elsewhere, the unlimited market for gold derives from its social role as money-commodity.

Few see any need to explain either the fixed-price situation, or why gold is always accepted on the market. Innes' view is that, 'the formation of the international gold standard ...formally established gold's role as the money-commodity' (Innes, op. cit., p. 45). He sees 'the international gold standard as the means by which currency exchange values *could become accurately known in terms of one another*' (ibid., p. 48, emph. ours.), hence the fixing of the price. This, unfortunately, says nothing about gold and makes its role dependent on the arbitrary whim of power politics. It immediately begs two questions: firstly, how is the central role of gold in international trade to be accounted for in the 5,000 or so years prior to the establishment of the international gold standard?; and secondly, why did its value remain more or less constant for centuries before the establishing of the gold standard?

The dramatic increase in the *value* of gold consequent upon the world supply changing its source from predominantly placer deposits to lode gold in the period c.1860-c.1880, threatened complete disruption of world trade, the medium of exchange being gold. But more fundamental still, the *measure* of value, which, by definition, is required to be constant, has changed the calibration of its own unit. One hour's labour-time no longer produced x amount of gold. Of what value, then, were those commodities the values of which were expressed in terms of that x amount? A great revolution in money itself was taking place. This had a third consequence: the reliability of gold as value-for-itself, i.e., as store of value, was now compromised. As value-for-itself, no value stored in this form any longer represented the same value as when they first left circulation. The socially necessary labour-time for the production of a unit of gold has changed, thereby affecting the value of all existing units.

The international gold standard, far from being the great institution of stability, as it is so often lauded in the literature, was a symptom of the instability which had befallen the unit of the value measure. Fixing the *price* of gold to a value reigning prior to this change, was an attempt to hold fast *de jure* that which had already been overturned *de facto*. Gold now had to be produced at the earlier, much lower, value, or not be produced at all.

The fixed price and the limitless market stand as two equal, but mutually independent conditions of gold economics. Neither the fixed price, nor the limitless market, are explained. Johnstone's approach is, therefore, empiricist. But Johnstone is, nevertheless, superior to Innes in that instead of trying to merge goldmining capital and other (specifically diamond digging) capital into a single kind of capital (Innes' 'monopoly capital'), he does recognise that monopoly, which did develop on the diggings, did *not* replicate itself on the Rand:

the different market conditions of gold compared with those of diamonds—specifically the different relationship between price and output—determined that the structure of ownership of the gold mines was to be less monopolistic than that of the diamond industry (Johnstone, op. cit., p. 14).

The problem lies only in Johnstone's accounting for this distinction. For him, the specific market price of diamonds was directly responsive to the relationship between supply and demand. Since realisation depended directly on market share, competition, under capital's own machinations, would tend to give way to monopoly. One capitalist, having thus captured control of supply, thus captures control of price. 'But this kind of control was not necessary in the gold industry', says Johnstone, 'because the price of gold was internationally fixed' (ibid., p. 14).

Here firstly, Johnstone confuses a controlled price with a fixed price. Monopoly in the diamond market stems not from the price being *fixed*, but from the price being *controlled*. He who controls the price can fix it at whatever level he wants. The goldmining capitalist had no control whatsoever over the gold price. Indeed, the fixed price controlled him, as Johnstone himself acknowledges above, 'the mining companies were unable to transfer increases in production costs to consumers in the form of price increases'. Put another way, when Johnstone talks of 'the different relationship between price and output' for diamonds and for gold, what is obscured is that for diamonds there was such a relationship, whereas for gold there was none.

Secondly, Johnstone again conflates the issue of a limited/unlimited market with that of a fixed price when he speaks of, 'the absence of any need to restrict output', in the context of a fixed price. This echoes his thinking on diamonds, where there was no need to restrict output in the context of a *controlled* price. While it is empirically true that 'what developed in the goldmining industry was an oligopolist but highly centralised structure of ownership and control', this does not follow from Johnstone's reasoning. Johnstone does not explain either why monopoly did not develop, nor why oligopoly did develop in the goldmining industry.

Johnstone, like Innes, lacks a theory. Neither of them are able to account for their observations. Johnstone's position as a scholar, however, is somewhat stronger than that of Innes. The former allows his material to suggest its own thesis, the latter imposes one from outside. Where Innes sees 'monopoly tendency towards "collusion"' (Innes, op. cit., p. 53), Johnstone describes the goldfields as being under the control of 'several large corporations' who

> came to establish common central organisations and to implement common measures in order to eliminate competition between the mining companies for factors of production, especially labour, and to rationalise the process of production (Johnstone, op. cit., p. 14).

This introduces the question of the form of competition in goldmining capital and the way in which this is manifested in the circuit of such capital.

Competition and the Group System

The existence of the Group System is generally understood in terms of three issues: firstly, collusion secondly, rationalisation; and thirdly, promotion. The first embraces especially the procurement of the factors of production (means of production and labour-power); the second entails streamlining of operations, amalgamations, centralisation of facilities and sharing of expertise; and the third involves competing for capital on the financial markets, reducing the risk to investments and representing the industry in dealings with the state.

While goldmining capitals do not compete to dispose of their product, they nevertheless do compete to procure the factors of production, and they do compete to secure capital on the financial markets. In the competition for the factors of production, there would clearly be a tendency towards monopsony. In the competition for finance, however, the role of monopoly/monopsony is different.

In the buying and selling of commodities in the market, monopoly/monopsony relates to attempts to gain control over price. In the financial markets, however, what is being peddled are not commodities, but rights to future shares of surplus-value in the form of money. The 'item' being offered for sale is therefore homogeneous across the entire landscape of the financial markets, regardless of the particular use-value from which the surplus-value is captured or to be captured.

In the money-market only lenders and borrowers face one another. The commodity has the same form—money. All specific forms of capital in accordance with its investment in particular spheres of production or circulation are here obliterated. It exists in the undifferentiated homogeneous form of independent value—money. The competition of individual spheres does not affect it (Marx, 1984, p. 368).

A monopoly in the stock market implies the control over *all* listed companies, regardless of the particularities of the commodities they respectively produce. This is impossible since it would quickly have to extend across all financial markets in pursuit of capital seeking outlets elsewhere. Such a situation would describe the sublation of all competition, and hence, capital.

But this does not mean that there are no benefits to be derived from a number of different capitals appearing on the financial markets as a single entity, or as a small number of highly colluding/co-operating entities. To begin with, they would enjoy greater prominence than were they to present themselves as single companies, thereby both attracting more capital, but also adding to the security of the investor. Secondly, any risk to the capital invested in them is spread across a number of productive units. This also works the other way in that the individual company needs not be quite as fearful of the withdrawal of its capital on account of not being able to extract sufficient surplus-value.

So, although there would be no point in striving to attain monopoly (or monopsony), there is distinct advantage to be had from some form of collective representation. Such collective representation would be relatively free from internal strain since no competitive advantage attaches to dissension. This does, however, assume that such companies are all competing for capital in the same financial markets. The paradox is that, on the one hand, when goldmining capitals compete for the same social capital, i.e., when their market share should be under threat, they stand more to gain from co-operating with, rather than undermining, one another. On the other hand, when they do not compete for the same social capital, when they pose no threat to one another's finance, they benefit from undermining one another without restraint.

Although there was a great deal of movement of capital around the world in the late nineteenth century, the financial markets were, nevertheless, relatively far less integrated than they are today, retaining much of a nationalist character. This was both in terms of their listings, and in terms of their 'reputations'. Although the City of London was the world's most developed financial centre, it was not necessarily the most secure. Berlin, e.g., with its joint-stock banks directly integrated with productive capital, did have a reputation for being very secure.[6] This, too, would be reflected in the competition between the Witwatersrand goldmining companies, based, as they were, in both centres.

Both benefits of putting up a united front on the financial markets — the attractiveness of large-scale representation and that of reduced risk — can be seen to be present in the Group System operating in the South African goldmining industry. These are discussed in the literature. Johnstone sees the significance of the Groups as lying in the provision of capital, the stabilisation of investment and in the rationalisation of production (Johnstone, op. cit., pp. 14-17). 'They... were able to provide amounts of capital and a continuity of investment which would have been beyond the resources of the average single goldmining company'. But at the same time:

> Given such factors as the Groups' large assets, their extensive involvement in the industry in terms of both investment and production, and their established position in the international capital markets, these shares were easily dealt with in these markets (ibid., p. 15).

The stabilisation of investment, which is synonymous with the minimisation of risk, again, worked both from the side of the investor, and from the side of the mine. Observes Johnstone, 'there was an improved chance that poor areas will be offset by good, and that even development will be maintained' (quoted in ibid., p. 15). To this end serves also the pursuit of rational production, under which falls the active promotion of company amalgamations, 'which the Groups were able to implement as they wished since they controlled the companies'. Large-scale production was, 'also the means of equalising the output of gold and preventing violent fluctuations in the price of the company's shares' (quoted in ibid., p. 16). Innes sees it similarly:

> Investment in goldmining had to be made as attractive as possible to the overseas investor which meant, in the final analysis, promising high returns at low risk. The former could be achieved by maintaining a low cost structure and the latter through spreading the risk over a number of ventures. The group system emerged as the organizational form through which both of these objectives could be achieved. By merging a number of

mining-finance houses together into distinctive 'groups' and thus bringing a large number of producing mines... under a single control, the Rand capitalists were able to spread the risk of investment in mining considerably (Innes, op. cit., p. 54).

It is clear that, even if not monopoly/monopsony, then oligopoly/oligopsony did develop in those areas where the goldmining companies did find themselves subject to competition. Neither Johnstone nor Innes, though, explain why the organisation of the industry should have stopped short of outright monopoly, as in Kimberley, and stabilised at oligops*ony*. We have already shown that it is inappropriate to talk of monopoly in goldmining, whether in selling the product or selling the shares, even if Innes insists on calling it that. Wheatcroft does make this distinction:

> the problem was diametrically opposed to Kimberley's. The magnates, however, very quickly saw that in the case of diamonds the price was variable and had fluctuated wildly in the two decades before amalgamation. The price of gold by contrast was fixed, and the problem was to control variable costs. In each case the answer was combination or cartelisation with opposite ends in view. In Kimberley the object was monopoly in the strict sense, a market condition with a single seller who can then dictate the price: upwards. On the Rand the object was monopsony, the market condition when there is a single buyer who can dictate the price: downwards. This was the aim towards which the magnates conspired in the critical matter of labour...,

and then proceeds to makes the crucial observation that:

> It was not necessary to amalgamate the mining companies if there was another institution. What De Beers Consolidated was to Kimberley, the Chamber of Mines was to Johannesburg (Wheatcroft, G., 1986, p. 132).

But the very existence of the Chamber of Mines requires explanation. This organ directly undermined the opportunities for the mining companies to compete with one another in the one area where classical competition remained possible: competition for factors of production. Goldmining capital might have been a peculiar kind of capital, but it was nevertheless still capital. As has been shown above, the product is always sold. It is only a matter of actually getting the material out of the ground. Any company unable to do so loses its capital and must dispose of its mining claim. Neighbouring capitals would be waiting for just such eventuality in the hope that they might be able to extract the gold themselves. There is, after all, only so much gold beneath each claim. This is exactly what underlies the reluctance of, e.g., J. B. Robinson & Co., to co-operate with the Chamber of Mines.

Numerous attempts were made to standardise the wages for black miners, each time derailed by some mines, in particular those of Robinson, breaking ranks by paying more, 'as a means of cutting their competitors' throats' (ibid., p. 131) and Albu, who, says Wheatcroft, 'was doing his best to raise black wages by not adhering to the scale already agreed upon; not of course from charitable motives, but to beggar his neighbours' (ibid., p. 186). In terms of the agreement which the Chamber reached with the mine managers in August 1890, that black wages be set at a maximum of £2 per month, wages were by October 1891 only reduced to £2 8s. 10d. The severe labour shortage of 1891, though, 'much as companies and mine managers wanted to reduce costs', led to several companies breaking ranks from the agreement as, 'they still more wanted to keep their mines milling' (ibid., p. 131). Robinson eventually left the Chamber to form a rival organisation, the Association of Mines, which held back even oligopoly. Innes describes the events:

> When the Chamber was formed in 1889 all the leading mining-finance houses, with the exception of J. B. Robinson and Co., joined it. Robinson held out alone against the Chamber until 1895 when, supported by two other large companies which had previously been members of the Chamber (Ad[olf] Goerz and Co. and G. and L. Albu and Co.), he formed a rival combination, the Association of Mines. In 1898, however, Goerz and Albu were enticed back into the Chamber and the subsequent collapse of the Association forced Robinson into the Chamber shortly afterwards (Innes, op. cit., p. 72n9).[7]

Although this information is provided by Innes, he skims lightly over it, failing to draw out its significance. He over-stresses the 'monopoly' character at the expense of the competition between capitals. This is sometimes done at the expense of factual accuracy. Although, e.g., the Chamber of Mines was formed in 1889, before the change-over to deep-level mining (1890-95), Innes could still say that 'the various groups had combined to form the Chamber of

Mines' in response to the wage increases consequent upon the increased labour demands of deep-level mining. In the light of the abandonment of the Chamber by some of its most powerful members, as described by Innes above, it is strange to then read that although the mining companies 'had to abandon their wage determination, ...they did not abandon the organisation they had established (the Chamber of Mines) to defend their combined interests' (ibid., p. 58). The *a priori* thesis of 'monopoly capitalism' stands between its protagonist and the obvious competition between capitals. Where such competition cannot be ignored, we read that monopoly capital, 'lacked the means of *enforcing* a common policy in the industry', and of, 'the difficulties monopoly capital faced in integrating itself' (ibid., p. 59). What, then, is the substance of this monopoly?

Although Robinson, Albu, Goerz, etc. were not competing against other goldmining capitals for disposal of their product, neither were they altogether competing in the same financial markets. The advantages of a single unified representation in the City of London was of less importance to those capitals partially based in continental markets, and of no importance to those based entirely elsewhere. Wernher Beit, Adolf Goerz, and G. and L. Albu were such companies. In addition to the idiosyncratic Robinson, these were the very companies in a position to defend their independence from the Chamber.

> The capital needs and structures of the several financial groups engaged in goldmining operations on the Rand varied considerably. The Werner, Beit complex had already access to a variety of dependable sources of continental capital which it used carefully and very successfully ...Unlike ...all the other financial groups, it [Gold Fields] was dependent on the small British investor. But neither it nor Werner Beit were in 1895 as vulnerable financially as their main competitors. ...The Albu and Goerz interests, backed by German banks, and therefore possessed of a more secure capital base, were in 1895 just in the process of becoming firmly established... (Kubicek, R., 1972, pp. 101-2).

It must immediately be said, though, that Wernher-Beit, their 'dependable sources of continental capital' notwithstanding, was nevertheless a London-based company.[8] When the great split came, in April 1896 in the aftermath of the disastrous Jameson Raid in which Beit was implicated, Wernher-Beit remained in the

Chamber while the twenty-two companies led by Robinson, the Albu brothers and Goerz seceded to form the Association of Mines (ibid., p. 196). One area of competition, though, viz., that for factors of production, remained. And it was in order to compete in this area that, in particular, J. B. Robinson, *inter alia*, kept his distance from the Chamber.

The general evening-out of strengths and weaknesses amongst the different mining companies which stems from their membership of the Groups, and, in particular, of the Chamber of Mines, has the effect of integrating them more closely together. Simultaneously, the presence of competition in the procurement of labour-power and means of production has the effect of rigidifying them in their autonomy. The one tendency directly undermines the other. The more they cohere, the more they atomise, and *vice versa*. The good of capital in general and that of capital in particular, which are mutually exclusive under conditions of competition, are here both present in each particular capital. They are each and every one torn between answering each of these contradictory calls — to cohere, and to atomise.

This can be seen as a particular expression of a general contradiction in all productive capital: that of the tendency towards proportionate production, i.e., the scientific and rational distribution and application of resources (as made explicit in social capital) against the tendency to override proportional production in the quest for surplus-value. In the Grundrisse, Marx puts it thus:

> *Proportionate production* ...only when it is capital's tendency to distribute itself in correct proportions, but equally its necessary tendency—since it strives limitlessly for surplus labour, surplus productivity, surplus consumption, etc.—to drive beyond the proportion. (In *competition* this inner tendency of capital appears as a compulsion exercised over it by *alien capital*, which drives it forward beyond the correct proportion... (Marx, 1973, p. 413, emph. orig.).

Albu, Goerz, etc., partly facilitated by the non-coincidence of their financial markets with that of their competitors, represent the rise to prominence of the atomising tendency. The control which the Groups exercise over the individual mines, and the way in which the central organisations came to be indispensable to those mines, meant that the elimination of competition between them for factors of production, represents the eventual dominance of the unifying tendency. 'It is found in practice that the real power is vested in the hands of that Group which has a sufficiently large share-holding to permit of its assuming control of the company. ...[The Groups were] ...formed for the purpose of financing, directing and controlling mining companies' (quoted in Johnstone, op. cit., p. 15).

The gold mines remain individually listed companies in their own right, but each 'contracts to one of the mining houses to provide various services to the company', the most important of these services being finance. But these also include consulting engineers, technicians, research laboratories, bulk purchase, legal support and secretarial services (Wilson, F., 1972, p. 22-3). Their character as particular capital is constrained by their relation to the Group (while they all, to a greater extent than would be the case for capital taking the full brunt of competition, assume some of the character of capital in general). As Johnstone puts it, 'they were owned by share-holders, and effectively controlled by the groups through majority share-holding' (Johnstone, op. cit., p. 15).

The companies, the groups and the shareholders

It has been argued throughout that the condition of a fixed-price commodity provides the incentive to set up production on a rational basis: elimination of uncertainty, flexible allocation of resources and integrated production. This is reinforced by the assurance that the product will be disposed of. This same point is made by Innes:

> The specific nature of the international market for gold had contradictory implications for the future of the development of the industry. On the one hand, conditions of an unlimited market at a fixed price meant that the gold industry would not be hampered as others, and especially the diamond industry, had been by market fluctuations and restrictions: the market for gold was unlimited and therefore the threat of over-production did not exist, while the existence of a fixed price meant that capitalists could *plan production* on the basis of *assured returns* on the sale of the commodity (Innes, op. cit., pp. 48-9, emph. ours).

But this can only be pursued in so far as it remains compatible with capital. It is not that, 'the specific nature of the international market for gold had contradictory implications for the future of the development of the industry', but that the contradiction which normally posits itself as a relation *between* particular capitals, as competition, here manifests itself *within* particular capital. The device contrived by capital to try to resolve this contradiction is what came to be known as the Group System.

Uncertainty in the goldmining industry obtains in three areas: (i) factors of production (ii) finance and (iii) the production process itself. Although the particular strategy in each area had to be different, they could, nonetheless all be addressed by the device of the Group System. In order for uncertainty to be eliminated, competition itself had to be eliminated. Note also that this is more than merely a question of reducing the costs, although cost-reduction can and will be discussed in respect of all three of these areas. Cost-reduction is, of itself, but a particular form of the rational allocation of resources. For capital, though, their coincidence is not automatic. Cost-reduction and planned production will therefore be discussed separately.

Cost-reduction

(i) Factors of production: Here the problem was both the extraordinarily high cost of the means of production, and the high cost of labour-power. The early goldmining industry had a curious overlap between the two. The high cost of means of production reflected the backwardness of South African economic conditions at the time. This same economic backwardness also accounted for the high value of white labour-power when compared, say, to the United Kingdom. This is taken up in detail below.

This condition did not apply to the value of black labour-power since (i) this was reproduced independently of the mainstream South African economy, and (ii) the value of black labour-power was, in any case, a great deal lower than that of white labour-power. But further, the mines did provide food and accommodation to black workers. These items, in company books, came under 'stores', and as such, formed part of the means of production.

As far as the high cost of means of production was concerned, the mines therefore had to address themselves to costs in the South African economy in general. Whatever advantage the mines might have gained from the relatively low value of black labour-power at the time quickly came to naught as this labour was also in extremely short supply. The market price of black labour, thus, ended up being extremely high, driven hence by competition between the mining companies. In respect of labour, it had to solve the problem of supply.

The Chamber of Mines served for both these purposes. The Chamber campaigned vigorously and endlessly to get the cost of means of production reduced. Its chief headaches in this area were the dynamite monopoly and the extortionate railway rates. The Annual Reports of the Chamber at the time are replete with letter after letter to the Government pleading for an end to the dynamite monopoly. So serious was the Chamber about getting the railway rates reduced that it repeatedly urged the state to expropriate the Netherlands South Africa Railway Company. Whenever Commissions of Enquiry were appointed, the Chamber prepared lavish studies to demonstrate the deleterious effect of South African costs generally on the conduct of its operations. The Chamber also purchased materials, stores and equipment in bulk for its members, thereby reducing the cost.

In respect of labour the strategy was two-fold: firstly, to encourage labour onto the market; secondly, to discourage competition for labour. The first involved setting up its own recruiting organisation in order to by-pass the exorbitant private recruiters, and to urge the state to more rigorously enforce the pass law in the rural areas. The pass law, indeed, was introduced on the initiative of the Chamber. Accommodating black workers in compounds and providing them with food allowed for both of these to enter the worker's consumption cheaper than had the mines to pay workers to buy these items individually.

The second part of the strategy called on all employers to agree on a maximum wage. This was attempted several times but success eluded the Chamber until into the twentieth century. The Chamber also deprecated the practice of mines poaching properly-contracted labour from one another, which was widespread on the Rand at the time. To this end the Chamber campaigned for and got a special pass system for the Witwatersrand, according to which workers were individually identified by a metal tag and hence, at least theoretically, traceable to their contracted employers.

(ii) Finance: The profit portion of surplus-value produced in joint-stock companies has to be divided between that accruing to the 'individual' who initiated the company, and the dividends to outside shareholders. The greater the ratio of public shares to those of directors, etc., the greater must be the portion of the surplus-value leaving the company. Dividends, together with interest payable on bank loans, constitute the cost of capital to the company. The more it can draw on internal sources of capital, the less would be its total capital cost.

The Group system of share-ownership has done much to reduce the cost of capital to the gold mines, by internalising what would otherwise have been external sources of capital. With reference to Figure 2, instead of the individual companies trading all their shares on the stock market, a portion (sufficiently large to ensure control) is subscribed by the Finance/Mining House. The latter then issues its own shares to the public, alongside the company now doing so with those of its shares not subscribed by the Finance/Mining House. The Finance/Mining House may establish its own new companies in which case shares in these companies would first be offered internally to its own member companies before these are issued to the public. The companies themselves, therefore, also hold shares in one another.

The Finance/Mining House also came to offer loans to its associated companies (Frankel, S., 1967, pp. 18, 21). One Finance/Mining House together with its associated companies constitute a 'Group'. Interface with the financial markets, and hence the cost of capital, is thus kept to a minimum. Even though this describes the ownership relationships which came to characterise the goldmining industry, these were less developed during the period considered by our thesis. Direct issues of shares to the public was more prevalent, or shares were issued simultaneously to the vendor and the public (ibid., p. 21).

(iii) Production process: Here the principal contributor to cost was the existence of a disaggregated multitude of autonomous production units: the mining companies. Not only did this mean the duplication of costly equipment and services, but it also made actual production operations less economical than they might otherwise have been. The Groups, through their control of the mining companies, could and did force through amalgamations which reduced overall working costs. The British investing public always found observers ready to extol the virtues of scale to them. The *Investor's Guardian,* under the heading, 'Advantages of Amalgamation', describes the following immediate benefits:

> With larger areas of mining ground larger bodies of ore can be developed, thus avoiding the necessity of the constant removals of the labour force from one portion of the mine to the other for the purposes of equalisation of yield. By the concentration of working underground, costs are reduced as to stoping, shovelling and tramming to the lowest limit consistent with efficiency. Large blocks of low grade ore are thus brought within the limit

of profitable treatment, thereby prolonging the life of the mines. These amalgamations, too, warrant the erection and running of additional and heavier stamps, and thus enhance the profit-earning capacity of the properties.

...Then as to the economy which amalgamation will secure in the saving of salaries of various mine employes, this will always be considerable even when duplication only is avoided (*Investor's Guardian*, 1910, p. 13).

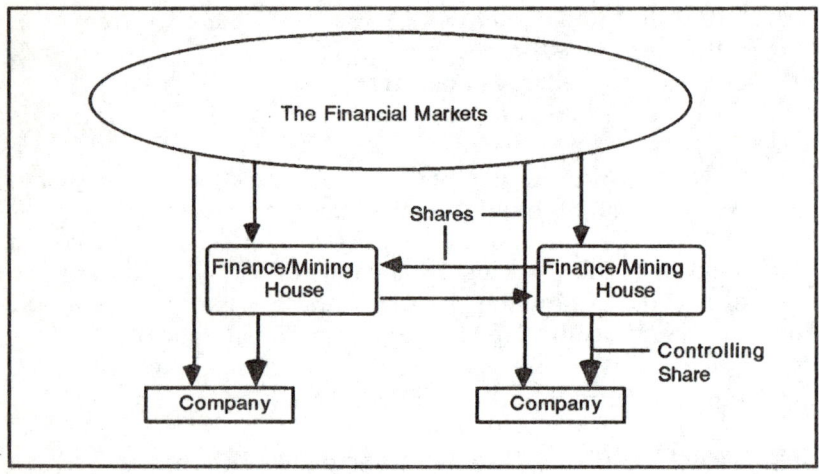

Figure 2 Share-ownership structure of the group system in Witwatersrand goldmining

The Groups have established various specialist central organisations which perform otherwise expensive tasks for all the associated companies. These specialists are engaged in a relentless pursuit of cheaper ways of doing things.

Towards planned production

(i) *Factors of production*: In order to set the procurement of the factors of production on a scientific basis, this has to be, firstly, centrally organised, secondly, of stable cost and thirdly, of stable supply. The centralised organisation of buying has already been mentioned. Centralised buying also has the effect of stabilising demand, both in terms of its volume, and in terms of its regularity. The stabilisation of costs proved to be a more intractable problem.

(ii) Finance: Finance in general, as already discussed above, is notorious for its unreliability. This is much more so the case with mining in particular. Frankel describes the problem thus:

> All mining enterprises display a considerable number of economic factors which particularly affect the mobility of capital investment. Mining enterprises are usually required to comply with fiscal and legal provisions (often subject to unpredictable changes) which determine the right of access to the mineral deposits and the conditions under which they may be worked. Some of these may lead to near monopoly conditions. Knowledge concerning the nature of the deposits, their value, and the prospective returns they may yield is often very imperfect. Such knowledge as may exist may not always be available to investors. Investment of capital on mineral exploration or development may be spread over long periods before any return at all can be contemplated. Once so embarked capital cannot be withdrawn without considerable or even total loss. There may be calls for additional capital investment without which the original investment cannot be made productive; such further capital investment may not at terms be forthcoming on suitable terms, owing to a change in political or economic circumstances which could not have been foreseen (Frankel, op. cit., p. 17).

But in terms of mining in general, goldmining finance (and especially Witwatersrand goldmining finance) tends to be more stable. This notwithstanding, it is in finance that the real power of the Groups as rational allocators of resources emerge. The individual companies do not in any significant way interface with the financial markets at all (Fig. 3). As absorbers of capital they loose their stand as individuals for part of a larger whole, 'a complex of capital-raising and risk-spreading institutions' (ibid., p. 18). The Group is free to allocate Capital invested by the public in the Finance/Mining House wherever in the complex it is most needed, thereby evening out the unevennesses within the Group. Companies can weather production shortfalls since there is no direct link between the company's performance and the market. Whomsoever invests in a Finance/Mining House will find his capital spread across a number of enterprises.

This reduces the risk to the investor and makes the investment more stable. On the basis of this more stable investment the Group is better able to plan its production and make longer time-horizons. This became even more effective with the amalgamation of whole groups into larger groups, allowing one Finance/Mining House out of each of the new agglomerations to emerge as the 'parent' company. 'The group system itself has undergone a steady process of integration, in as much as the number of groups has declined, their scope has become larger and their interrelations have become closer' (quoted in Innes, op. cit., p. 57). Thus did the mining companies come to be represented in the financial markets by fewer groups, each with vastly increased resources at its disposal. Rationalisation of production could thereby be that more effectively pursued (ibid., p. 56).

(iii) Production process: Amalgamation, as discussed above, was not only a cost-saving device, but also a rationalising one. This took the form of integrating production. Most of the cadastral boundaries on the Witwatersrand cut the Reef transversely, causing the ore body to lie partly under one property and partly under another. With mines becoming increasingly deep a point must be reached beyond which it would simply be cheaper to dig horizontally underground (galleries), than to dig two shafts from the surface. Similarly, new mines can be planned in terms of the attributes of the ore body, rather than in terms of cadastral considerations.

To an extent these amalgamations were forced onto the mining companies by the revolutions in gold processing taking place in the last decade of the nineteenth century. Mining was no longer simply a hole in the ground. An increasingly large portion of the production operations were beginning to take place above ground, first in the physical pulverisation of the ore (milling), but also, and more importantly, in the chemical processing of the millings. Huge cyanidation plants became an essential part of the production process after the pyretic ores were struck in 1890-1.

Few individual mines could either afford, or had the scale of operations to warrant such an installation. But this, again, is not so straightforward, as Frankel points out:

> At any given time the life of a mine is determined by the capacity of its plant in relation to the total ore which can be economically mined. The size of the plant will, in turn, depend on the capital invested. There can be an increase in both life and

scale of operations only if the grade is lowered (Frankel, op. cit., p. 10).[9]

Every single cheapening device, every rationalising measure, in whatever area of goldmining expenditure, had a direct bearing on how much of the proven gold was payable. The mines were therefore not only concerned with the cheapening of labour costs and the rationalisation of the supply, they were concerned with lowering *all* costs and rationalising *all* operations. It is, of course, true that labour-costs made up about fifty percent of total costs (see any WCM or TCM Annual Report), but the object was to attain both lower costs and planned production, and not just lower labour costs. It is an error to see the Group System as merely a connivance against labour, as is so often the case in the literature.

Such company amalgamations also allowed a rationalisation of labour allocation within the new Group. Sudden increases and drops in labour demands at particular sites could be responded to by physically shifting gangs of workers to where they were needed, rather than by modifying output.

Once an expansion in the scale of operations has been embarked upon, it was hardly possible for the mines to avoid their evolution into a tightly-integrated, all-seeing, all-knowing, entity. Upgrading the scale of production involves new capital investment which in turn involves new variables for the investor — a new range of uncertainties. Frankel again:

> An increase in the scale of operations requiring new capital investment for additional plant, etc., really involves the creation of a new mine in place of the old. Any change in the expectations which have led to such capital investment creates a new situation—either there will have been over-investment of capital and capital losses will result, or the calculations will again have to be revised and further new capital investment take place, resulting in a different relation between the life, scale of operations and grade (Frankel, S., 1936, p. 552).

What then remains is that the difference between the different mining companies reduces itself to the bounty bestowed on it by nature. They all have roughly the same technology and expertise and capital available to extract their gold. Of course, unevenness must exist since, e.g., some mines are older than others, or the physical conditions within the mines do not always allow for the

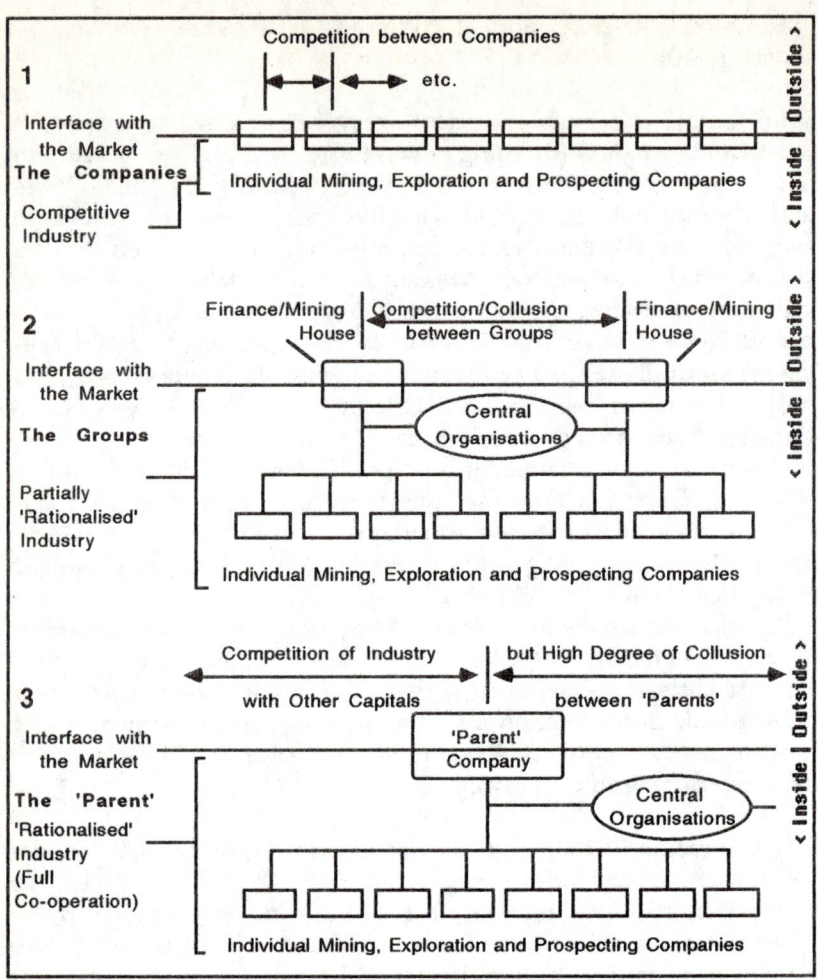

Figure 3	Competition and the group system in early Witwatersrand goldmining
Sources:	Compiled from Frankel, S., 1967, pp. 18, 20; Houghton, D., 1973, p. 103; Innes, D., 1984, pp. 52-7; Johnstone, F., 1976, pp. 14-17; Wheatcroft, G., 1986, passim; Wilson, F., 1972, pp. 22-4.

application of the same technology. All competition between them is, for all intents and purposes, eliminated. What matters between them is their respective contributions to the Group. What matters to the financial markets, is the production of the Group as a whole.

In the financial markets, and, indeed, in the labour and equipment and stores, etc., markets the Groups, at this level an oligopoly, compete against all *outside* the industry. The competition and atomisation which would have existed at the level of the individual goldmining companies is transferred to that of the Groups together, while the rationality and balancing of needs with resources of the financial markets, i.e., of social capital, is extended to the individual companies.

But the mining companies were never wholly-owned subsidiaries either, for the finance houses themselves needed to maintain their position as intermediaries between the financial markets and the mining companies themselves, which is how they claim their share of the total surplus-value. The example of Corner House serves to illustrate the general principle:

> Each mine was floated as a joint-stock company with its own directors and its own manager. The Corner House kept control through share ownership but more importantly by dominating the board of directors. These mining companies were never what would later be called wholly-owned subsidiaries. That, indeed, would defeat the whole object of the exercise which was to raise capital by flotation, by the sale of vendor's interest, in some cases by 'booming' the stock (Wheatcroft, op. cit., p. 133).

The Group System of organisation is goldmining capital's way of keeping one step ahead of its investors, but by the same token, it keeps ahead of its own inner nature as capital.

Competition for factors of production

From the fact that the price of gold on the world market had, during the period under review, been fixed, most of the literature on the political economy of South African goldmining deduce nothing more than that high production costs could not be passed on to the consumer and hence, somehow, had to be reduced. The conditions of Witwatersrand mining allowed very few options for achieving this and the only real opening for cost-reduction lay in the cost of labour. Hence the need for an extra-cheap labour supply. Some take this a bit further by saying that labour costs could be reduced in one of a number of ways: either by making the supply so large that it bids itself down; by reducing the demand to a single buyer thus able to dictate the price; or by rapidly developing the productivity of labour through a massive mechanisation programme. Of those writers who get this far, most argue that the first two were implemented and the third not. The point here is that most writers ascribe some kind of special significance to the fact that gold could only be disposed of at a fixed price, but usually that significance is taken to extend no further than the need for a cheap labour force.

Goldmining capital, with the exchange value of its commodity constant, would want the value of its c, too, to be constant. It therefore puts in place an industry planned down to the worker's butcher's account. The need or constancy in wage costs is obscured by a number of other constraints. The first was the high *market price* of black labour against its value and the high *value* of white labour (irrespective of its price). Black wages had to be reduced to be more consistent with its value and the value of white labour had to be reduced in order that white wages, too, could be reduced. This important problematic is obscured in the literature by the excessive concentration on wage struggles and strikes in the industry at the expense of examining the question of value.

The mining industry did not entirely have its way with white labour, but it did with black labour, reflecting the different relationships which capital had with the two components of its labour force. For a whole host of reasons which lie partly *outside* of the capital-labour relation, the cost of white labour could not be kept constant. With black labour, though, the Chamber had great success. Indeed, black mine wages did not only draw comment for their extremely low level, but also for their *constancy*. This is confirmed by different writers. Hence, e.g., Katzen:

> Besides being low, the wages of African miners are extremely stable. 'Nothing has changed so little in South Africa as the black man's rate of pay'. Nowhere is this more true than in the goldmining industry. African miners' wages display almost no cyclical sensitivity at all (Katzen, L., 1964, p. 22),

or Wilson, 'during the first eighty years of their [the gold mines'] existence the real wages of black miners did not increase at all; indeed, over the period 1889 to 1969, they seem actually to have fallen' (Wilson, op. cit., p. 141).

Keeping the value of labour-power *constant* is quite a different matter to that of keeping the value of labour-power *low*. The first has to do with proportional production; the second with surplus-value extraction. These contradictory tendencies are present in all productive capital, but it is the competitive market which constantly shifts capital's attention from one to the other, as the imbalance becomes clear to it from its failure to effectively compete. Where the product is not disposed of through competition, both of these tendencies come to fall within the conscious domain of capital.

The Chamber of Mines, as collective representative of goldmining capital, did not escape these contradictory tendencies now existing side-by-side within all of its members. Indeed, the contradiction became explicit in the Chamber in that its members came to express the drive for surplus-value, hence the threat of competition, while the Chamber itself came to express the drive for proportional production. Although the Chamber of Mines was established as early as 1889, it can be seen from the following description of its brief that opportunities for disunity were many:

> Since its formation in 1889 the Chamber has been mainly engaged in endeavouring to secure the mines relief from the heavy burdens imposed on them. The cost of living, and hence the rate of wages of the white employees, has been rendered unduly high by the duties on the necessaries of life and on all articles of ordinary requirement, while heavy railway tariffs and the dynamite monopoly, and the difficulty of obtaining an adequate supply of native labour at reasonable pay, have militated against the profitable working and development of the industry as a whole (Industrial Commission of Enquiry Report, 1897, p. iii).[10]

Most of the competition, in fact, occurred in the area of wages. This led not only to the frequent breaking of ranks by members after every agreement for uniform wages, but the eventual imposition of a pass law to deter members from encouraging each others' labourers to 'desert'. Contrary to the view often expressed in the literature, the pass law in early goldmining had as much to do with controlling capital, as it had to do with controlling labour. The Chamber collected extremely detailed breakdowns of every company's labour requirements and utilisation on a month-by-month basis. It approached governments, not only in southern Africa, but as far afield as Hungary and Japan, with proposals for labour importation.

By setting up its central recruiting organisation, the Witwatersrand Native Labour Association, the Chamber took the task of recruiting out of the hands of the individual companies. It thereby assumed centralised control over the actual proletarianisation process (although it was reluctant to do so and tried throughout to get the state to assume this task). The point, though, is that the actual proletarianisation process was then set on a scientific basis. The drive towards rational production drove capital all the way to creating its own labour force to its own specifications. Or, to use modern informal parlance, the black mine workforce was a 'designer workforce'. The mines clearly showed that abstract labour was not in the interests of the mining industry. Its labour-allocation system approaches that of planning. But it was planning within the confines of capital: it was bound to be contradictory.

Competition for finance

It is clear from the Witwatersrand Chamber of Mines Annual Reports (WCM), that the question of dividends and the mines' difficulty in paying them did not weigh lightly on managements' minds (WCM, 1889 to WCM, 1899).[11] The mines, of course, had all the normal problems of joint-stock companies: competition for shares; size of dividends; proportion of preference to ordinary shares; pressure from the securities markets (currency markets as such did not yet exist, although gold and currency speculation was already a problem for capital by 1873);[12] the balancing of internally-financed expansion with, on the one hand, the magnitude/regularity of dividends, and with, on the other hand, new issues; etc.

The Witwatersrand gold mines had two advantages over ordinary joint-stock companies: (i) they dealt in a product for which there was always a demand, and although the price was fixed, i.e., it could not be raised, it could also not be lowered. The security of value-for-itself is thereby extended into fictitious capital riding on its back. Serious shareholders would be inclined to exercise greater patience with goldmining companies than with others; and (ii) the Witwatersrand goldmining companies made a dramatic entry onto the London Stock Exchange in the midst of the huge number of flotations in 1895-6,[13] of which an extraordinarily large number were mining companies.[14] A cursory glance through the list of newly registered mining companies shows the overwhelming number of them to have been goldmining companies, and of these, by far the larger number to be located on the Witwatersrand (see London Stock Exchange Yearbooks for 1896, 1897 and 1898).

This helped to ensure for the mines a collective reputation which tended to obscure the performance of the individual mining company. This was further reinforced already from 1887 with the appearance in the City of London of the mining finance houses promoting their 'members' in the financial markets. 'The first financial house was formed in 1887, to bring the Rand goldfields before the British public' (Wilson, op cit., p. 23). Subsequently, even if the stock market was generally depressed, a good word could always be found for 'Kaffirs', as Witwatersrand gold shares were affectionately referred to:

> Since the passing of the deep depression which was for so many years the chief characteristic of the Kaffir market there have been many bright intervals but these have been no[t long] ...lasting. ...These varying moods of the market... do not at all reflect the nature of the management of the mines there. If it had been otherwise, I should have seen the control at Johannesburg listless and careless at one time and energetic and painstaking at another, whereas I have invariably found those in authority ever on the alert to procure the utmost production from their great undertakings at the lowest cost (*Investor's Guardian*, 1910, p. 5).

Goldmining shares in general and Wits gold shares in particular tended to be more secure than they otherwise might have been. This does not mean that they were immune to the general vicissitudes of the stock market. Indeed, the Rand had its shares of disastrous speculative ventures, and paid dearly for them. According to one account, The Coronation Syndicate, 'was a speculative venture. The collapse of the Syndicate, shares in which were "boomed" in Europe, did much to discredit Rand mining shares generally' (Fraser, M. and Jeeves, A., 1977, p. 137n27). Neither does it mean that goldmining shares could not collapse, as, indeed, they did in 1889 and 1895 (Marks, S. and Trapido, S., 1979, p. 51), and again in 1902 (Richardson, P., 1977, p. 88). It is merely being suggested that investors would tend to persevere longer with goldmining shares than they would with other shares, and that goldmining shares would enjoy some cushioning in bad times from the abandonment of non-goldmining shares for goldmining shares. The problem of dividend payments could therefore have been a lot worse for the mines than was actually the case.

However, amongst themselves the individual companies had no particular advantage one over the other in the eyes of the shareholder. Most shareholders were not serious holders and did not have the time to concern themselves with the mechanics of goldmining, so that the sound reasons for a delay in payment tended to be irrelevant.[15] Pressure was therefore on the *individual* mining companies to pay quickly and pay well as share capital was less likely to redistribute itself amongst the companies than it was to leave the Witwatersrand altogether. The description of the mining companies competing amongst themselves for their share of the 'Kaffirs' is therefore a doubly apt one.

The technical contest with nature

In most of the literature on South African goldmining the assertion is made that goldmining is a backward sector unable to mechanise or take advantage of scientific or technical developments in general. For some, this would be on account of financial constraints while for others it would be because Marx said so. To those who have actually studied the goldmining industry empirically the exact opposite is clearly the case.

The fixed price of the product on the market is not the only constraint that goldmining capital has to contend with. There are constraints of a more technical kind, the most important of which is the grade of the ore, i.e., the quantity of gold per unit weight of ore. The cost of production of a gramme of gold is clearly seriously affected by whether that gramme was retrieved from one or two tonnes of ore. In the latter case it would not only cost more to retrieve it, but also take longer to do so. While both of these are critical, the literature discusses only the implications of cost but not that of time.

To deal with cost first: one gramme or even half a gramme per ton can make a difference as to whether a particular ore body can be profitably mined or not, i.e., whether it is payable. So much so that the term 'payability' has become part of the goldmining lexicon, warranting whole studies on the subject.[16] Non-payability means that the ore remains in the ground. It is therefore in the interest of goldmining capital, firstly, to be able to know beforehand whether a particular ore body is payable, and if so, the extent of that payability. The higher the payability, the smaller can be the mass of the ore body which will yield the minimum quantity of gold which will pay for its own extraction. This can make the difference between whether a mine is developed or not. The mines have therefore made every effort to develop and apply the geo-sciences (geology, geo-chemistry and geo-physics — especially the latter, for its remote-sensing capabilities).

Secondly, payability is not synonymous with grade. Grade indicates the quantity of gold *present* in the ore, while payability states the quantity of gold *retrievable* from the ore. This is not the same thing. E.g., there might be quite a lot of gold in a particular body of ore, but the ore is of such a nature that it will either not yield the gold at all, or only yield it at prohibitive expense. In this case a process has to be found, together with the necessary plant and equipment, which will not only force the gold from the ore, but do so at a cost which will still deliver the gold beneath the fixed market price. One situation like this, which almost brought the mining industry to its knees, was the encounter of pyretic ore in 1890. The invention of the MacArthur-Forrest cyanide process did not come a moment too soon and was adopted with great urgency by the industry.

The situation confronting competitive capital is that when the cost of producing the product is too high, it will not *sell*. In the case of non-competitive capital, the product will simply not be *produced*. Mines which do not produce do not pay dividends. Mines which do not pay dividends loose their capital. The pressure is therefore on every mine manager to bring every gramme of gold under his property within payability. For this reason he will pursue every technical device and every development in technique which promises to do this. One of the most striking features of Rand goldmining right from the start, and attested to by many, is the enthusiasm with which the mines introduced new machinery. Hence, e.g.,

> it can be said that in no other mining region has money been spent more lavishly on labour-saving devices than on the Rand. Many new forms of machinery for this purpose have been invented and developed here, and any new methods which have been invented abroad are not long allowed to remain untried in these fields (quoted in Richardson, P. and Van Helten, J.-J., 1982, p. 84, and referring to the turn of the century),

and,

> The severe competition which have now been proceeding for many months in the mines of the various Witwatersrand companies, cannot fail to prove of marked advantage to the industry in bringing well to the front those machines which will

thus have convincingly demonstrated their superiority over their rivals (*Investor's Guardian*, 1910, p. 10).

The mines also, through their organisation the Chamber of Mines, right from inception followed patent applications very closely, not hesitating to challenge those which threaten to lumber them with royalty charges for processes already used by them (see especially WCMs for the 1890s). This does not mean that the mines sought to restrain innovation in general, quite the contrary. They were simply operating on extremely narrow cost margins.

The time constraint derives from two imperatives: one technical; the other financial. Mines have to take a decision as early on in the planning/development process on which ore bodies or sections of ore bodies are worth exploiting and which are not. The fundamental consideration, of course, is the grade of the ore, i.e., how much gold is actually present in it. The next question is then whether that ore can be profitably extracted. On the basis of this, such decisions are made as to where to sink the shaft, how many cross cuts to make and at which levels to make them, etc. Such developments are costly in themselves and often unalterable once carried through. If it then turns out that the ore is not as payable as anticipated on the surface, it must be left in place until technique allows for its payable extraction but the capital invested would then be idle. If technique develops to such an extent as to bring an ore body judged unpayable and by-passed by the superstructure into payability, such an ore body could be out of reach forever. It is therefore critical to the mines to know of new techniques and innovations as soon as they are announced, and to implement them as quickly as possible.

The second consideration of time is financial. A mine spending years digging its way past non-payable ore in order to reach payable ore is unable to pay dividends to its shareholders, who are all the time aware of neighbouring mines on the same listings which do pay dividends. In goldmining, unlike in industries competing through the price of their products, backward technique cannot be concealed and protected behind price manipulation. The item is either produced or it is not produced.

Conclusion

What, then, remains of competition amongst mining capital? The mines do not compete for labour: this is provided to all the mines at a fixed cost. They do not outdo each other in labour-productivity, or in their ability to extract the gold from nature, since they pool technique and engineering services and constantly inform each other of more efficient methods of extraction. Competition such as there is in the financial markets is reduced to a minimum since the Rand attracts capital as a collectivity. Once the gold appears on the market, it is all of the same quality. The difference in their individual claims to their share of the profits, thus, reduces itself to the *quantity* of gold which each has managed to produce. From each according to his product, to each according to his product.

Regardless of whether or not consciously motivated, the much-remarked-on co-operation and sharing which characterises Rand goldmining must not only have helped to promote the conception of the Rand as a collectivity in the financial markets while blurring the individuals, but also, and more importantly for this study, have had a dampening effect on the competition for capital amongst the companies. But in addition to this, such co-operation and sharing was consciously motivated.

The whole thrust of the industry's efforts during this period has been directed at the elimination of competition between its members, which had the effect of enhancing the Rand gold mines as a singularity in competition with all others in the financial markets.

It is not often appreciated by scholarship that gold has a social role quite distinct from its role as an ordinary commodity. Of the works consulted, only Williams and Trewhela start from the particular social role of gold. Both these scholars are taken up for critique elsewhere in these pages. In his book, *The Politics of Race*, Hillel Ticktin denies any special role for gold as money-commodity in explaining South Africa, but nevertheless admits that 'the main export of South Africa is gold' and that 'it is the money-commodity itself' (op. cit., pp. 78-9). In saying that 'South Africa cannot ...be explained only through the peculiarities of the money-commodity', he does, though, imply some role. It remains for Ticktin to show, or at least suggest, just what this role might be and how it related to the remainder of the explanation of South Africa. But he does not say why this is so and offers no critique of those theses which argue the contrary. Ticktin does begin to suggest the direction of his thinking when he says that, 'gold has ...been countercyclical and helped to stabilise the class which held the gold mines'. The 'class which held the gold mines' happens to be the most powerful class in South Africa. Ticktin does not say whether the counter-cyclical character of gold derives from its role as money-commodity or from something else. Even if one saw the peculiarities of the money-commodity in explaining South Africa as merely subordinate to some more fundamental explanation, then the nature of that subordination needs to be laid bare. A real political economy of gold still remains to be developed. The present essay, we hope, might contribute to such a theory.

Goldmining capital extracts surplus-value from labour-power as do all other capitals. The profits, however, are not distributed by capturing surplus-value from each other at the point of sale of the product in the market, for they pay the same wages, employ the same technology, have, broadly speaking, the same productivity of labour and they all sell the product at the same price. The gold mines, in the period under investigation, paid a portion of the surplus-value out as dividends, and the remainder ended up divided amongst themselves according to the share of each in the total product.

Competition between them therefore took a form more akin to emulation, with each trying to extract more from nature than the other. The extent to which they could scientifically pursue this was constrained only by their relation to their shareholders. All other constraints they have continually striven to overcome, all chance and uncertainty they have sought to abolish, impelled by the lack of competition to dispose of the product. The picture which emerges is somewhat different to that offered by some authors, of the Randlords as uncontrollable gamblers, whose striving for unity was motivated by caprice:

> The behaviour of generations of miners, speculators and hoarders testifies to the peculiar fascination of gold, to which the flint-eyed captains of Rand finance were far from immune. The same individuals who might cheerfully have left tons of, say, uneconomic copper in a mine dedicated themselves on the Rand to the single minded pursuit of the last possible pennyweight of gold. Thus was the propensity to gamble intensified. Nevertheless, the mine owners sought to hedge their bets and so to protect themselves from the various uncertainties to which they were vulnerable. The group system of control was one device for accomplishing this (Jeeves, A., 1985, p. 7).

Johnstone is correct when he says that there was no need for monopoly on the Witwatersrand. He further says that, 'what developed in the goldmining industry was an oligopolist but highly centralised structure of ownership and control' (Johnstone, op. cit., p. 14). One might also go so far as to say that this is the most scientific production attainable within the confines of capital. This is, however, quite different from describing the Group System as 'a modern structure of capitalist enterprise', as does Johnstone (ibid., p. 16). If this were so, then modern capitalism has had plenty of time to replicate it elsewhere. But this has not happened. We would argue that it is the most modern structure attainable for a capitalist enterprise, but only *provided that it is a gold-producing enterprise*, for this structure follows from conditions only applicable to gold-production.

Notes

1. With the extension of credit economy across the full landscape of economic relations, as in the post-1971 world, this role assumes a different form.
2. Our thesis is not concerned with the manner in which monopolies in general extract their share of the total surplus-value. We do, however, offer some ideas on this for goldmining capital.
3. For Marx on the historical development of competition out of monopoly, see Marx, 1973, pp. 649-52. Monopoly in relation to rent has been discussed in Section I, above.
4. A dramatic modern form of this is take-over for the purpose of asset-stripping, on the one side, and *kanban*, on the other.
5. 'As soon as it begins to sense itself and become conscious of itself as a barrier to development, it seeks refuge in forms which, by restricting free competition, seem to make the rule of capital more perfect...' (Marx, 1973, p. 651).
6. The weight of German capital was sufficiently important for even Lionel Phillips to be concerned over its possible withdrawal (see letter to FitzPatrick, 12 June 1897, quoted in Fraser, M. and Jeeves, A., 1977, p. 105).
7. In 1892 this 'monopoly' consisted of 67 separate companies (ibid., p. 72n12).
8. 'By 1895 the most important financial group on the Rand ...traded under the modest name of H. Eckstein & Co., ...but was always known as the Corner House. The company was controlled from London, where the name of the firm was

...Wernher, Beit & Co.' (quoted in Wheatcroft, G., 1986, p. 8).

9 By, 'if the grade is lowered', he means, 'if the payability threshold of the grade is lowered'.

10 (hereafter ICE). The *Report of the Industrial Commission of Enquiry into the Gold-mining Industry* (otherwise known as the *Burger Report*) of 1897 is a remarkable document. It was commissioned by the ZAR Government to report on the then depression in the goldmining industry and was published by the Chamber of Mines. It provides meticulous details of every aspect of Witwatersrand gold mining at the time, including a detailed comparison with goldmining around the world. We could find the Report mentioned only half a dozen times in the literature. That this critical document remains virtually unknown, and where known so little-explored, is to the serious detriment of scholarship.

11 E.g., '...Sooner than pay thousands of pounds in wages, and be unable to make a return to our shareholders, we will close down the mines' (evidence of George Albu, mine owner, to the Commission, ICE, p. 14). WCM and TCM refer to the Annual Reports of the Witwatersrand Chamber of Mines and the later-renamed Transvaal Chamber of Mines.

12 Speculation, as a capitalist's principal activity, was at this stage still strongly distasteful to 'respectable' international bankers, who, during the difficult year of 1873, watched the movement of the gold premium with some trepidation (see, e.g., the correspondence of August Belmont and Co. of New York to N. M. Rothschild and Sons of London (virtually all letters of 1873, but especially between 2 January 1873 and 27 March 1873, Rothschild Archives London, XI/62/23A). The monetary panic of 1873 was at its height between early September and early October. During the panic the free market price of gold consistently declined in New York. In other words, gold was not called upon to play its role as money as it was already capital and, at that point, less lucrative than other forms of capital. Investments flowed into Greenbacks, banks special loans, etc.

13 'Moreover, 1896 is entered upon with such large stores of capital seeking employment that no improvement in the money market is yet in sight', this lack of improvement 'in the money market' referring to the relatively depressed

years of 1894-5 (see Preface to the London Stock Exchange Yearbook, 1896).

14 'As regards the expansion of both 1894 and 1895, it is noticeable that the majority of the companies have been for the purpose of acquiring and working mining properties' (ibid.).

15 'It may be said that three-quarters of all transactions in mining shares come under the heading of gambling, pure and simple. The average gambler in mines cares nothing for the intrinsic value of a property; he does not study its past history, he does not intend to hold the shares for dividends—in fact, he wants to know nothing about the mine. All he asks for is a large profit on his speculation in a minimum period...' (Curle, J., 1902, p. 337).

16 This may be exemplified by the title of a 1937 book by J. Gray on the subject of the Wits fields: *Payable Gold: An Intimate Record of the History of the Discovery of the Payable Witwatersrand Gold Fields and of Johannesburg in 1886 and 1887*; Johannesburg.

7 The black goldminers and the 'fixed gold price'

Introduction

The landscape of South African economic history is characterised by many extremely complex interrelationships. Ultimately, the situation reduces itself to the relationship between capital and labour, and most scholars accept this. However, the particular expression of this relationship in South Africa, as in any other particular situation, can only be grasped in its particularity by an 'appropriation of the material in detail', to use Marx's phrase. The complexity of this relationship, particularly as regards the black goldmining workforce, is reflected in South African scholarship by the diverse variety of positions defended, rather than by the depth and richness of explanation offered.

A question which remains far from resolved is that of the genesis, evolution and character of the South African proletariat. There is no argument that the South African working class is not the same kind of unitary entity as, say, the British working class. There is, however, debate about whether it is one working class split in two, or two distinct working classes and whether the great divide lies between migrant and settled workers, black and white workers, skilled and unskilled workers, organised and unorganised workers, or goldmining workers and the rest. The black workforce on the gold mines have been variously characterised: at the one extreme, by Paul Trewhela, as 'the proletariat [which] could not be a stranger to the notion of *civil war*, and *revolution*' (Trewhela, P., 1986a, p. 19, emph. orig.), to the other by the Unity Movement, who 'did not consider the mining proletariat a genuine part of the working class' (Ticktin, op. cit., p. 11). Whatever the view, the black goldmining workforce tends to get singled out from the rest. Most scholars seek to explain the peculiarities of the goldmining workforce by the upheavals and transformations which swept southern Africa in the wake of the opening-up of the diamond and gold fields, especially the latter.

Although the present essay is not concerned with seeking answers to all these questions, it does offer a reassessment of the proletarianisation process which followed in the wake of the great mineral discoveries and suggests an alternative explanation for the peculiarities of the black labour force of the goldmining industry. Here we are particularly interested in the level of wages, the oppressive form of housing, and the control on movement with which that workforce came to be associated. We will show that these peculiarities flow from the *manner* in which the black workforce was created. This manner reflects three projects on the part of capital: (i) to reduce the market price of black labour-power; (ii) to ascertain the value of black labour-power; and (iii) to control the value of black labour-power. We will address the creation of the black mine workforce out of the economies extant in southern Africa at the time, and the forces working against such creation. The proposition is that, during the period when goldmining was the only significant industry in the country, the development of the social homogenisation of labour was foreclosed by the particular manner in which the black population was proletarianised. The question of abstract labour, as conceived by a number of prominent theorists, will be examined.

This examination covers (more or less) the period opening with the land dispossessions around the diamond diggings in Kimberley to the establishment of closed compounds for black goldminers on the Witwatersrand, i.e., c.1875-c.1907. The notion that the migrant workers in general and mine workers in particular have 'access to the means of production' in the labour Reserves has been an albatross around the necks of all sociology-based analyses (but not only these). The relationship which evolved between the labour Reserves and capital after the Land Acts of 1913, 1927 and 1936 falls beyond the scope of our study. We confine ourselves only to the establishment of a workforce adequate to the production of gold under the conditions flowing out of the changeover from placers to lodes. The relationship between the goldmining industry, manufacturing and agriculture will, therefore, not be addressed here.

'Abstract labour' and South African scholarship

In the literature examining the peculiarities of the black workforce on the South African gold mines, we could find only three studies in which the category of abstract labour serves as an explanatory device. These are an article by Michael Williams, *South African Capitalism — Neo-Ricardianism or Marxism?* (Williams, op. cit.), a manuscript forming part of the unpublished doctoral thesis of Paul Trewhela entitled *The Problem of Subject in the Modern World: The Role of the Proletariat of Southern Africa* (Trewhela, P., 1986a) and Hillel Ticktin's book *The Politics of Race* (Ticktin, op. cit.) Trewhela draws on Williams and the latter, in turn, bases himself on the first two chapters of Karl Marx's *Capital,* Volume I (Marx, 1983). Both Williams and Trewhela argue that goldmining labour constitutes abstract labour in view of the social character of gold as the money commodity. Ticktin, on the other hand, sees abstract labour as that condition of labour which renders it ideal for industrial capitalism, viz., generalised fluidity and homogeneity. We shall discuss the Williams-Trewhela conception first, then proceed to Ticktin.

Central to Williams' conception is the idea that the homogeneity of the material gold approximates the homogeneity of abstract labour more adequately than that of any other substance. He then makes the subtle point that, 'in this way, *gold-digging,* a specific form of concrete labour, becomes the *medium* of expressing its opposite, human labour in the abstract'. Since gold serves for the expression of the values of all commodities, proceeds Williams, 'gold, as it comes out of the South African mines, "is forthwith the direct incarnation of all human labour"' (Williams, M., 1975, p. 7, emph. orig. The words between double quotation marks are those of Marx quoted by Williams).

Trewhela picks up from here, arguing that the *concrete* labour of gold-digging *is* abstract labour, and as such, it *is* all other labour. A number of categories are here conflated. The first is that simple labour is conflated with abstract labour. Trewhela talks of the, 'appropriation of abstract, homogeneous, simple human labour in the form of gold-digging'. It is not our intention here to open an extensive discussion of the two-fold character of labour, merely to examine it to the extent that it has a bearing on goldmining labour in particular.[1]

The term abstract labour is not without its difficulties. Firstly, *abstract labour* needs to be distinguished from *human labour in the abstract.* The former is an aspect of the character of labour when its concreteness can only become social through a mediation. Labour assumes the character of abstract labour only under particular conditions, viz., that of generalised commodity production and, in particular, wage-labour. Human labour in the abstract, as opposed to abstract labour, refers simply to the total quantum of social labour, abstracted from its aggregated specificities, which an economy, any economy, has at its disposal for allocation to its various necessary productive tasks. Hence, wherever there is human labour, there will be human labour in the abstract, but not necessarily abstract labour.[2]

Labour brought to activity by capital has a two-fold character: (i) it is, like all labour in every form of economy, *concrete* labour. That is to say, it is a physical[3] activity involving a distinct series of physical actions which can, on the basis of informed anticipation or past experience, be expected to result in a specific product the material character of which results from that specific labour and no other.

Concrete labour can be either complex or simple. Concrete labour is labour as it is performed by the individual. It is therefore also private labour. Complex labour expresses the concentration within the individual of a great deal of skill, and conjointly, the lack of such skill in machines, i.e., socially. This was a feature of labour during the early part of capitalism and, indeed, in more primitive modes of production. Simple labour indicates the lack of such skill in the individual, but not necessarily, however, its social lack. At the dawn of civilisation such social lack would be the case.

In modern society skill is socially at the same time both concentrated and diffused in machines. Concentrated because it can be executed more efficiently through machines, and diffused because the number of individuals in society capable of executing this skill is greatly increased — the individual being required to dispose only over labour sufficiently simple to enable his operation of the machine.[4] The complex labour-simple labour polarity is therefore a private labour-social labour polarity. This, however, talks only of the *technical* nature of labour and says nothing yet of its social *form*.

(ii) Regardless of how simple or complex concrete labour might be, it would have to be engaged under particular social conditions in order for it to assume the character of abstract labour. Commodity-producing labour is rendered social through the act of exchange. Exchange lends it the capacity to convert itself from one particular concrete labour to any other particular concrete labour. Labour exercised under these conditions is, in addition to its specific concrete character, also abstract labour. Without its exchange it remains merely potential social labour. It is therefore exercised with this intention. When such intention is generalised across an economy, then all labour is exercised for its abstractness, rather than its concreteness. The aim of wage-labour is to objectify itself in a form which encapsulates that abstractness.

From the side of labour, wage-labour is labour for money. From the side of capital, wage-labour is continuously driven beyond the bounds of its own particularity. Concrete labour becomes increasingly simple and in its simplicity increasingly comes to approximate abstract labour. With the development of industrial production, skill loses its particularity (both in its specificity and in its concentration in the particular individual) to become generalised social skills applicable to all production. This merely expresses the process by which concrete labour comes increasingly to approximate abstract labour. Industrial production itself propels concrete labour in this way. This is Ticktin's point, viz., 'the social reduction of labour to a common form' (loc. cit.).

The increasing homogenisation of *concrete* labour does not, however, render concrete labour *abstract*. If concrete labours were to become so homogenised that one unit of concrete labour becomes indistinguishable from another unit of concrete labour, such concrete labour will not have attained a condition of abstract labour. Rather, abstract labour will have been sublated, and concrete labour would come to express, directly, human labour in the abstract.

Williams makes the observation that, 'the reason why gold has come to serve as the money material is because its *natural* substance approximates more closely than any other commodity to the *social* content of commodities, to wit, abstract, undifferentiated and therefore equal human labour' (Williams, M., 1975, p. 9, emph. orig.). It is true that this social form of production, which, like all production, must apportion its total social labour-time, its *human labour in the abstract*, to its various necessary production tasks. In a form of economy where all concrete labours assume the character of abstract labour, such human labour in the abstract must first exist as abstract labour in order to be so apportioned.

Since the measure of human labour in the abstract is labour-time, labour-time is the measure of all labour. The unit of measure is a fixed time during which labour of the minimum skill valid as labour is exercised. Abstract labour, however, being a particular form of labour mediated *through exchange,* already announces that the unit of labour-time requires a bearer to reliably transfer measurement from one concrete embodiment into another. It requires a medium of circulation. Commodity economy itself sets aside a portion of its human labour in the abstract for the purpose of producing this bearer. Such production is part of its necessary labour tasks.

For such bearer, such circulating medium, to have any bearing capacity, it must, like the commodities between which it mediates, be a product of human labour and reducible to its unit, x labour-time. But it is concrete labour which is so reducible and as such, it shares the two-fold character of all commodity-producing labour: that of being at the same time concrete labour and abstract labour. The product of this concrete labour, the material gold, is as homogeneous, divisible and re-unitable as the labour which produced it is *in its character as abstract labour*. This concrete labour becomes the medium for expressing not only its own character as abstract labour, but, since abstract labour is 'undifferentiated and therefore equal human labour', for expressing *all* abstract labour. The product of such concrete labour is called, according to its concrete character, gold; according to its abstract character, money. Human labour in the abstract, i.e., the total body of available social labour, is, in commodity economy, thus allocated in the units in which abstract labour is allocated, viz., in units of money, unit weights of gold.

When Williams, then, says that, *'gold-digging,* a specific form of concrete labour, becomes the *medium* for expressing its opposite, human labour in the abstract' (ibid, emph. orig.), he is saying a number of things, some of which are correct and some of which are not. Gold digging, of course, is a form of concrete labour. This labour is also a medium, and that medium does express the opposite of concrete labour. But the opposite of concrete labour is not 'human labour in the abstract', but rather, abstract labour.[5] It is such a medium only under conditions of exchange, and ceases to be so outside of exchange. Since human labour in the abstract requires, under conditions of commodity economy, to be expressed as abstract labour, this concrete labour *becomes* the medium for expressing abstract labour. Williams' deduction is based on the following passage by Marx:

> The body of the commodity that serves as the equivalent, figures as the materialisation of human labour in the abstract, and is at the same time the product of some specifically useful concrete labour. This concrete labour becomes, therefore, the medium for expressing abstract human labour (Marx, 1983, p. 64).

When Marx makes a distinction in this passage, between 'human labour in the abstract' and 'abstract human labour', he is actually talking about two different things, and not merely being editorial with himself, as Williams apparently thinks. Williams uses 'human labour in the abstract' where Marx uses 'abstract human labour', thereby conflating what Marx keeps clearly separate.

Williams then argues further that, 'gold, as it comes out of the South African mines "is forthwith the direct incarnation of all human labour"' (loc. cit., p. 177). It must already follow from the argument advanced above that gold is the *indirect* incarnation of all human labour. It is true, though, that it *appears* to be the direct incarnation of all human labour. But this appearance is none other than a moment of the very appearance according to which the social relation between the products of concrete labours presents itself as an attribute of their material selves. Marx, in Volume I of his *Capital*, makes the point thus:

> We followed up this *false appearance* to its final establishment, which is complete so soon as the universal equivalent form becomes identified with the bodily form of a particular commodity, and thus crystallised into the money-form. What *appears to happen* is, not that gold becomes money, in consequence of all other commodities expressing their values in it, but, on the contrary, that all other commodities universally express their values in gold, because it is money. The intermediate steps of the process vanish in the result and leave no trace behind. Commodities find their own value already completely represented, without any initiative on their part, in another commodity existing in company with them. These objects, gold and silver, just as they come out of the bowels of the earth, are forthwith the direct incarnation of all human labour. Hence the magic of money (Marx, 1983, pp. 95-6, emph. ours).

Williams, by removing from this general argument (and adapting) the one sentence: '...gold and silver, just as they come out of the bowels of the earth, are forthwith the direct incarnation of all human labour', manages to give it a meaning exactly opposite to that intended by its author. Far from gold being, 'the direct incarnation of all human labour', it is all human labour which *becomes identified with* gold. Marx, in his *Grundrisse*, puts it thus:

In the form of exchange value, labour-time is required to become objectified in a commodity which expresses no more than its quota or quantity, which is indifferent to its own natural properties, and which can therefore be metamorphosed into— i.e., exchanged for—every other commodity which objectifies the same labour-time. The object should have this character of generality, which contradicts its natural particularity. This contradiction can be overcome only by objectifying it: i.e., by positing the commodity in a double form, first in its natural immediate form, then in its mediated form, as money. The latter is possible only because a particular commodity becomes, as it were, the general substance of exchange values, or because the exchange values of commodities *becomes identified with* a particular commodity different from all the others (Marx, 1973, p. 168, emph. ours).

It is on these misreadings of Marx by Williams that Trewhela basis his own argument. Trewhela argues, following Williams, that gold-producing labour, since it produces abstract wealth, has the *concrete* character of abstract labour. It is argued that the product of goldmining represents not itself, but the entire spectrum of products of concrete labours, i.e., products in the abstract; the *concrete* labour of gold digging *is* abstract labour. For Trewhela, the value form renders the labouring activity of gold-digging, 'immediately abstract, immediately social, homogeneous human labour'. Thus Trewhela:

> ...abstract, homogeneous, qualitatively equal human labour in the immediately useful form of gold-digging; ...the form by which the concrete labour of the collective gold producer is, by the very nature of the value form, immediately abstract, immediately social, homogeneous human labour (Trewhela, P., 1986, p. 8).

Here use-value II is conflated with use-value I. Concrete labour produces only use-value I, in this case the material gold. Use-value II comes about by the value relation, 'the very nature of the value form', that arises between different instances of use-value I. All labour is compared to gold-producing labour because gold is the universal equivalent. But this is not the same as gold-producing labour = abstract labour, for in such equation even gold-producing labour is but a part of the pool of universal human labour and thus void of attributes, and hence figuring as a *relative*, rather than an equivalent expression of value. If gold-producing labour were *actually* this attributeless universal human labour, were identical with it, then *all* labour must be *actually* gold-producing labour, whether the product of that labour is shoes, caviar or gold. Gold as use-value II merely *officiates* as abstract labour; it does not *become* abstract labour.

> The *body* of the commodity that serves as the equivalent, figures as the materialisation of human labour in the abstract, and is at the same time the product of some specifically useful concrete labour. This concrete labour becomes, therefore, *the medium* for expressing abstract human labour (Marx, 1983, p. 64, emph. ours).

Gold producing labour is immediately social, but not for the reasons offered by Trewhela. It is immediately social (use-value II), but not because it is abstract; on the contrary. It is immediately social *because it is concrete*. It is immediately social not as use-value I — for as such it is a commodity and by definition not immediately social — but as use-value II, which is not a commodity and is hence immediately social. Abstract labour is abstract precisely because it is abstracted from this or that 'immediately useful form'. Precisely because the labour here discussed comes in 'the immediately useful form of gold-digging', it cannot be abstract. It is different from all other labours in the same way that any particular labour is different from all other labours. Or, as Marx put it, 'digging gold, mining iron, cultivating wheat and weaving silk are qualitatively different kinds of labour' (Marx, 1977b, p. 29). In other words, there is nothing exceptional about gold-digging labour as concrete labour. To talk of gold digging as 'homogeneous, qualitatively equal human labour', is to say nothing more than that gold-digging is qualitatively equal to gold-digging.

Gold-digging is further described as, '...labour unmystified by any particular concrete form or skill, labour whose particularity lies solely in its general lack of particularity' (Trewhela, P., 1986a, p. 8). If labour is ever mystified, it is not by its 'concrete form', but by its social form, for the concrete form produces a use-value and, 'the mystical character of commodities does not originate... in their use-values' (Marx, 1983, p. 76). Labour which does not come in 'any particular concrete form' cannot produce a use-value. Gold is a use-value, the particular concrete form of the labour which produced it being gold digging. Indeed, it is only as use-value, as a product of specific concrete labour, that anything can play the role of equivalent.

The 'particularity' of gold digging is gold digging. The gold digger produces neither abstract wealth nor money, or even the money material, but a metallic substance called gold. That a metal is capable of being mined does not presuppose particular *social relations*, like relations of exchange, which posit abstract labour, but only certain *technical capacities*. What labour claims from nature is the substance of gold, which becomes the use-value gold. This may or may not take the social form of a commodity which, in turn, may or may not take the higher social form of money. That this metallic substance does become money presupposes its being a commodity which, in turn, presupposes its being a use-value.

The implications of the Williams-Trewhela thesis is far from exhausted. For example, if the concrete labour which produces the money-commodity has the single-fold character of abstract labour alone, then, firstly, what labour is it an abstraction of?; and, secondly, so too then, must cattle-rearing labour, shell-cleaning labour, mat-weaving labour, woman-raising labour, etc., have the single-fold character of abstract labour, for cattle, shells, mats, women, etc., were all, at some time or other, the money-commodity. Although these have since ceased to be money (some have even ceased to be commodities), the concrete labour producing them persists. If, then, with gold as money-commodity, gold-digging labour is *in its immediate concrete character* actually abstract labour, then either gold was always money and always will be money, or there was no gold-digging prior to gold assuming the social role of money-commodity, i.e, it was never a commodity or even a product.

It is one of the recurring errors in scholarship to consider the particular use-value and the universal use-value of gold in isolation from each other, i.e, as if there is no relation between them. When gold is regarded in its role as money, it is forgotten in its role as commodity. When examined as commodity, its role as money disappears from view. One of the founding philosophers of socialism, Lenin, does not escape this charge when he suggests that when gold ceases to be the repository of value, it ceases to be a use-value (Lenin, V. I., 1921, pp. 113-4).[6]

At the same time, apart from the general point of the equation of simple labour with abstract labour, Trewhela does argue that the labour of goldmining is abstract labour on account of the social nature of the product, viz., money. In other words, looking at the relation from the vantage point of the product, gold is a supercommodity in that it is exchangeable against all other commodities. Gold is a commodity,

> whose useful character is solely that it serves as the repository of value as such; that, as against all other commodities which appear in competition as so many hostile particulars, this one commodity is at all times the soul of all and immediately convertible into all. To this degree, we have here a commodity standing outside of competition: the monarch of commodities, raised up by all the others as on a throne. ...it is their crowned head: it is the Messiah of commodities elevated over their own heads by all the rest, that living god in comparison with which all other commodities are merely mortals. In the circulation process, where values are realised, all other commodities must attain to the state of grace—gold alone is born to it (Trewhela, 1986a, p. 15).

The social form of labour as abstract labour expresses itself in the exercise of that labour as wage-labour. As wage-labour, labour-power is a commodity like any other. The labour which produced labour-power, likewise, has a two-fold character. It only becomes social upon alienation. Wage-labour is therefore, by definition, labour for money. The object of wage-labour is, therefore, an abstract commodity, money. It is not, however, correct to say that since gold is money, gold-producing labour is abstract labour. Gold, as use-value II, is produced for the *utility* of its exchangeability. Other commodities exchange against it for that precise reason. No commodity can be an exchange value without first being a use-value. In the case of gold, it is a use-value (use-value II) *because* it is an exchange value — the exchange value associated with use-value I. Without its *universal* exchangeability it would not be produced as use-value II. It is as this second use-value, use-value II, that it officiates as the equivalent expression of the value of *all* other commodities. If it were not a universal equivalent, but merely a general equivalent, it would do so as use-value I. The point here is that the value of a thing is always expressed in the use-value of another thing. The equivalent form of value, whether particular, general or universal, is necessarily a *use-value*. Hence Marx:

> With the equivalent form it is just the contrary. The very essence of this form is that the *material* commodity itself ...just as it is, expresses value, and is endowed with the form of value

by Nature itself. Of course, this holds good only so long as the value-relation exists... (Marx, 1983, p. 63, emph. ours).

Marx's last line, which reads, 'This holds good only so long as the value-relation exists', serves to bring Williams' fetishised conceptions of money, gold and gold-digging labour into sharp relief. If gold-digging labour actually is abstract labour, concrete abstract labour, so to speak, and gold, 'as it comes out of the South African mines is forthwith the direct incarnation of all human labour', then the conditionality of the value-relation falls away. The implication is that, regardless of the form of economy, gold-digging labour always has been and always will be abstract labour, while gold always has been and always will be money. This same value-relation is what labour cannot escape from so long as it remains wage-labour.

The value of labour-power

The question of the value of labour-power in South African goldmining is hardly ever addressed in the literature. It is indirectly discussed by Wolpe as 'the conditions of the production and the reproduction of labour-power' (Wolpe, H., 1972, p. 428), and by Williams as 'level of subsistence' (Williams, op. cit., p. 7). Somewhat arbitrarily, the phrase 'cost of production and reproduction of labour-power' is sometimes introduced to assist in explaining the labour Reserves. Ticktin does give the value of labour-power brief attention (Ticktin, op. cit., pp. 3, 13-15, 56-7, 74-7). It does not, though, appear amongst his, 'Four categories [which] have to be elaborated and understood in order to develop a cogent political economy of South Africa' (ibid., p. 8).

Williams discusses it at some length, unfortunately somewhat dogmatically (Williams, op. cit., pp. 6-7). The latter sees the value of labour-power as a function of wage-labour, rather than of labour-power. Trewhela makes only a few brief remarks about the value of labour-power (1986a, p. 17), but of the three South African scholars reviewed in detail, he is the most penetrating. Edna Bonacich, who writes within an orthodox economics paradigm, offers a most constructive examination of the value of labour-power and split labour markets. Her article will be reviewed separately elsewhere.[7] In the discussion which follows, much attention will be given to Williams and to Trewhela. This is for two reasons: firstly, they are the only scholars working on South Africa who have seriously tried to employ the value of labour-power as a category of their analysis, albeit in both cases a weak use of the category; and secondly, Trewhela remains unpublished while Williams' ideas appeared in an obscure journal.

The concept of the value of labour-power is not only wrongly understood, but also wrongly applied. It is suggested that this leads to an erroneous reading of the dynamic of the time. We conclude by suggesting that: (i) the value of labour-power was different for black workers from what it was for white workers, meaning automatically that wage levels would be different; (ii) far from black workers being underpaid, they were being, at least until the outbreak of the South African War (1899-1902), grossly overpaid; and (iii), ultimately, that the object of early goldmining capital was not to split black workers from white workers (this was already given), but rather to split black goldmining workers from black workers in general.

Irrespective of the social form in which labour-power is exercised, it remains labour-power, and as such requires the prior consumption of a certain sum of products (whatever social form these may take) in order to be produced. Capitalist commodity economy is not synonymous with economy, nor is wage-labour synonymous with labour. One can therefore talk of the value of labour-power without the exercise of that labour-power necessarily taking the form of wage-labour. The solitary hermit, the member of the Stone-age hunting band and the individual in communist society still need to consume a certain quantity of use-values in order to produce and reproduce their labour-power. All societies, whatever their form, need to consume a certain portion of their products in order to reproduce their total social labour from day to day, year to year and generation to generation.

> So far, therefore, as labour is a creator of use-value, is useful labour, it is a necessary condition, independent of all forms of society, for the existence of the human race; it is an external nature-imposed necessity, without which there can be no material exchanges between man and Nature, and therefore no life (Marx, 1983, p. 50).

It is sometimes assumed that the question of the value of labour-power, i.e., the necessary means of subsistence, arises only where the class of labourers forms a proletariat, or where a proletariat is in the process of formation. Williams holds that the value of labour-power is something to be determined by the class struggle.

Value of labour-power is equated with level of wages. These, we will show in a moment, are not the same thing.

Although Williams acknowledges that the value of labour-power embraces also the labourer's 'historically developed social needs, which become second nature' (op. cit., p. 7 quoting Marx), he confines this conception strictly to a proletariat and an already existing one. For this he draws on Marx when the latter says that means of subsistence also depends on, 'the conditions under which, the class of free labourers *has been formed*', (quoted in ibid., p. 7) adding the emphasis himself. If it is suggested that by this Marx means to say that the question of value of labour-power arises by labour-power being realised within the circulation process of capital, i.e., when labour-power is exercised as wage-labour, then it becomes very difficult to explain why Marx backs up this general point with the following footnote, 'Hence the Roman Villicus, as overlooker of the agricultural slaves, received "more meagre fare than working slaves, because his work was lighter"' (Marx, 1983, p. 168n1).

Williams then sees his task as having to 'ascertain... what was *to become* the *"historically developed* social needs" of the workers...' (Williams, op. cit., p. 7, emph. ours). This prediction of the past can only be understood if the 'historically developed' is deemed not yet to exist. And so, indeed, it is, for Williams goes on to talk of, 'the "degree of comfort" which was *to be established* as second nature' (ibid., p. 7, emph. ours). 'Second nature', i.e., the cumulative past concentrated in the present, is 'to be established' in the future.

On the basis of this temporal conflation Williams is then in a position to ask, 'Who is to determine what the level of subsistence of workers ought to be'? (ibid.) and to assert that, 'The mining capitalists were able to determine for themselves what the living standards of others ought to be only because they had prevented the working class from having a modicum of say in the matter' (ibid.).

Value of labour-power, skill, living standard and wage level

The literature seldom differentiates between value of labour-power, skill, standard of living and level of wages. What differentiation there is, is partial, in that skill is differentiated from level of wages. However, level of wages is often used interchangeably with value of labour-power. No attempt is made to isolate standard of living and the relationship between all of these is never explored. The importance of keeping these separate will become clear in the discussion of the relation between capital and black labour in the early goldmining industry. The idea for the moment is only to separate them conceptually.

We start with the distinction between the value of labour-power and level of wages. The former, as discussed above, describes the average socially necessary labour-time for the production of a unit labour-power. The wage describes the going market price of that unit. Price is the form in which value expresses itself. Its actual magnitude is thus necessarily imperfect, emerging under the influence of supply and demand as the market price. The value expressed in any particular market price describes a point along a path of oscillation around the actual price: now above, now below. With machine production and credit economy the relationship between market price and value becomes even more complex. Capital strives to keep the wage as close to the value of labour-power as possible, it being in its interests to pay neither above nor below that value. That value itself, though, might change.

As with all commodities, labour-power produced in environments with different productivities of labour have different values, taking unskilled labour as the unit. It is therefore possible for two groups of workers to be brought together from two different environments to work side by side, do identical work, and yet have different values of labour-power. It is not possible to say whether two groups of workers, each paid at different levels of wages, are better or worse off vis-à-vis each other, without considering also the respective values of their labour-power and the standards of their living. A given unit of value would represent different proportions of each of their necessary labour-times respectively.

It is possible: (i) for group A to be paid twice as much as group B, with group A getting poorer while group B grows very wealthy, if A is paid below its value and B above; (ii) for group A to have the same value of labour-power as group B, yet require twice as much in wages in order to maintain their respective standards of living, such as, e.g., unskilled labour in the countryside compared to unskilled labour in the city; and (iii) for groups A and B to have the same standard of living, but very different values of labour-power and hence levels of wages as, e.g., workers who go to work in an environment with a lower productivity of labour than the environment in which their colleagues remain behind. These distinctions are particularly important when examining the wages of black and white mine workers in early South African goldmining.

One of the problems with the fixed gold price is that it was fixed at a level which represented a magnitude of value far lower than the value needed to produce it under late-nineteenth century conditions. This is a consequence of the average social labour-time necessary for the production of a unit of gold having risen with the changeover from mainly placer to mainly lode mining on a world scale. For reasons internal to the workings of the international monetary system at the time, this change could not be allowed to feed through to the rate at which gold exchanged for the rest of commodities. Around the world, goldmining operations were either discontinued or never even started on account of the value of the labour-power being higher than could be accommodated within that represented by the fixed price of the gold to be retrieved by that labour-power (Lock, op. cit., pp. 44-5, 106, 511-13).

Enter South Africa

The mid-nineteenth century was the beginning of the age of generalised free wage-labour, i.e., the process of the abolition of slavery had, for all intents and purposes, been completed worldwide.[8] This means that the social form by which lode gold had been mined during the era of placer predominance, viz. working to death, was henceforth no longer available. Where transitional forms were still extant, these were used, e.g., indentured Chinese labour in South Africa. At the same time, the period also witnessed the generalisation of industrial production. Mining in industrialised countries benefited directly from this generalised industrialisation by being enabled to reduce their production costs. Mining in such areas could thus be restructured on the basis of *low-value, high productivity* labour-power, in other words, a labour force costing less but maintaining a high standard of living.

The value of Witwatersrand gold,[9] as it happened, exceeded that of placer gold to a much greater degree than elsewhere, not only on account of the comparatively poor grade of Wits gold (which is the only account mentioned in the literature), but also on account of the relative backwardness of the South African economy at the time. At the same time, the value of labour-power of indigenous African societies was a great deal lower than the world average for mining labour. There was therefore the prospect that this gold might be retrievable after all. One writer theorises that,

'If an ore body similar to South Africa's had been discovered in Australia, Canada, or the United States, it would most certainly have been left in the ground because of an inability to mobilize the right type of workforce' (Crush, J., 1991, p. 1). 'The right type of labour force', in the global context, would have been one of low-value and high-productivity. The backwardness of the South African economy did not allow for a high-productivity labour force. For mining capital to have turned the indigenous population into a workforce of general industrial productivity would have meant mining capital doing for South Africa in the space of a few years what all of previous history had done for Europe and North America: create an industrial society. For a low-value, high-productivity labour force to be created, the value of its labour-power would first have to increase (as it acquires industrial skills and improves its standard of living) alongside the increase in productivity, until full industrialisation begins to reduce the value of labour-power while continuing to increase the productivity of labour. Mining capital had to circumvent this bulge in the graph.

In 1901, a skilled miner with a wife and three children lived on £294 per annum on the Reef, while a similar family in Massachusetts lived on £170 per annum at a similar standard of living (TCM, 1902, Annexure, Exh. 15 and 15b). The value of labour-power on the Reef in 1897 is estimated to have been 25% higher than in California (Hall's evidence, ICE, 1897, p. 420). As an index based on Erith, England at 100, the cost of living in Scranton, N. America was 69, while at Johannesburg it was 128 (ICE, p. 158). The difference made by the general level of industrialisation of the environment in which the mine operates, may be illustrated by a comparison of the working costs of the Alaska Treadwell mine (AT) with that of Crown Reef (CR) on the Witwatersrand. AT's working costs in 1897 came to 4s. 8d. per ton of ore milled, to £1 6s. 8d. per ton for CR. At the same time the labour costs of AT came to 2s. 11d. per ton as against 15s. 6d. per ton for CR. The significance of mining in or close to an industrial environment is brought out by the fact that despite CR's paying six times more for labour than did AT, labour constituted a lower percentage of CR's total expenditure (57.98%) than it did in the case of AT (63.78%) (table in ICE, p. 212). AT's expenditure on the products of heavy industry is so much lower than that of CR, that its comparatively lower labour costs registers as a higher percentage of its total than does the labour costs of CR.

South African historiography consistently departs from the premise that South African goldmining required a cheap labour force on account of the low grade of its ore. Typical would be, e.g., the following observation:

> Profits are largely dependent on low production costs for two reasons. Firstly, because the average gold content of the ore is low. Secondly, the internationally controlled price of gold prevents the mining companies from transferring any increases in working costs to the consumers. Consequently within this narrowly circumscribed cost structure, the usual area of cost minimisation has been wages. The task then for the mine-owners has been to create and contain a vast supply of cheap African labour (Webster, E., 1978, p. 9).

This is true, but it is also misleading. The point is that the world over, gold production found itself in this position on account of the fixing of the price of gold to an inappropriate value. Cheap labour forces were needed everywhere for goldmining to continue. In South Africa, one happened to be available, subject to some slight modification. Over and above this comes the problem of the low grade of South African ore *as an extenuating factor*. It is in this global context, rather than that of 'imperialism' or 'monopoly capitalism', that the present essay seeks to understand the peculiarity which is South African goldmining.

This low-value labour force found in South Africa, though more than sufficient in number, was not a proletariat. The problem confronting capital was: How was this labour force to be proletarianised *without* raising the value of its labour-power? Or, put in a different way, could this labour force be brought to modern, industrial utility without making the kind of social investment that was historically made to produce an industrial workforce in Europe?

Since this labour, its numerical preponderance notwithstanding, did not readily offer itself in the market, it was scarce. On account of this scarcity, the different goldmining capitals competed with one another to attract sufficient labour, thereby bidding up its *price*. It was clear to capital right from the start that this labour was being paid *above* its value. Black labour, in the first decade or so of goldmining in South Africa, was being paid a 'wage' completely out of proportion to its 'frugal needs' (to use Legassick's phrase).

The concern of capital was, therefore, two-fold: firstly, to proletarianise this source of labour, thereby taking care of the supply; and secondly, to preserve as far as possible the relatively low value of this labour-power throughout that proletarianisation process and beyond. In Europe proletarianisation can be described as having been a 'natural' process in the sense that it sprang forth out of the dynamic of changes taking place within late medieval and mercantile economies. Capital itself paid nothing towards this process, indeed, its primitive accumulation was part of that process. In South Africa every worker had to be individually brought to the workplace at the expense of the individual capitalist seeking to engage him. With his own acute sense of what partial proletarianisation means, Lionel Phillips, in motivation of centralised recruiting, observed, 'The objection to allowing individual recruiting would not apply, if the whole world was open, as it does in a comparatively narrow field in Africa' (Phillips to Evans, 17, April 1903, quoted in Fraser, M. and Jeeves, A., 1977, pp. 119-120). Before the worker had even started earning a wage, he was already a cost item on the books. The cost to the gold mines of bringing a single black worker to the Reef ranged from £2 10s. to £3 2s. 6d., while their monthly wage ranged from £2 10s. to £3. In other words, each worker already cost the company one month's wages before he had even started working (ICE, pp. 14, 23).

That the labour of these labourers was unlikely to develop fully into free wage-labour was therefore inherent to the very manner in which proletarianisation took place. To entice or coerce individual workers to the workplace is one thing (and a costly thing, let it not be forgotten), but to create an entire class is quite another thing altogether. This was well beyond the capacity of capital, either individually or collectively. The state was continually called on by capital to fulfil this task. It was not always both able and willing to oblige, though it was often neither.

In an examination of the detailed interaction between the different natural economies in the southern African region, on the one hand, and the new capitalist economy in their midst, on the other, reveals a complete transformation of the African economies and the gradual (and uneven) metamorphosis of African natural labour into wage-labour. What emerges appears to be some way from the notion, so prevalent in leftwing literature, of African societies being laid waste by a rapacious capitalism right from the start. Those works which do try to examine this area anew include Bundy, C., 1979, Harris, P., 1982 and Richardson, P. and Van Helten, J.-J., 1982. They make for refreshing reading. We would go a little further than Bundy, when he argues that there was:

a substantially more positive response by African peasants to economic changes and market opportunities than is usually indicated; that an adapted form of the prevailing subsistence methods provided hundreds of thousands of Africans with a preferable alternative to wage labour on white colonists' terms in the form of limited participation in the produce market (Bundy, op. cit., p. 13).

As a relation between different forms of economy, the picture which emerges is one of trading the surplus living labour of the African economies for accumulated labour objectified in the plants and machinery of capitalism. What was surplus to the African societies was necessary to the capitalist one. It is, therefore, capitalist society which was in the weaker position. Unlike the African societies, it had nothing to dispose of without undermining itself. Which society gained and which lost by this transaction, cannot be straightaway deduced from the eventual triumph of one over the other. A case could be made that both the African societies and capitalism gained by this early relationship, since they both tapped into a resource which was otherwise unusable by either. To calculate this lies beyond the purpose of this paper. But if this were to be attempted, then, at the very least, reckoning would have to be made of the cycles of reproduction: (i) daily; (ii) annually; and (iii) from one generation to the next. Only the daily cycle was effected on the mines, and then only partially, i.e., while the worker was actually sojourning there. The annual and generational cycles were effected fully within the African economies. 'There are always other reasons why the natives went home—ploughing, sowing or getting married', declared George Albu to the ICE (loc. cit., p. 197). Production for all three cycles was therefore able to proceed uninterruptedly in addition to the *consumption* for a part of the daily cycle no longer being a deduction from that production. Regardless of whether the mine workers received any cash wages at all, their reproduction represented less of a drain on the African economies and hence a nett *increase* in their surplus product. Their tenacity in the face of capitalism was therefore much greater than the literature would suggest.

The history of the gold mines' black labour supply in the period up to the wake of the South African War can be periodised into two distinct phases: (i) a period of enticement (1886-96); and (ii) a period of coercion (1896-1903). The first period derives its character from the independence and relative prosperity of the various economies in which Africans lived.[10] These were not all the same. They ranged from patriarchal redistributive chiefdoms to peasant economies on the verge of capitalist agriculture.[11] But in all cases the factor determining their response to the enormous and rapidly-expanding demand for their produce and their labour, was the fact that in their traditional economies they had *labour to spare*. This, at bottom, is why they did not constitute a proletariat. Put differently, had this not been the case, their economies would have been undermined from the start. Instead, they prospered.

This prosperity has its direct reflection in the relationship between capital and labour.[12] During this phase of the relationship between mining capital and its black workforce, capital was on its best behaviour towards its 'Kaffirs'. Management, e.g., was ever mindful not to give Africans a bad impression of the Witwatersrand and took great care not to offend African tribal sensitivities.

> The desirability of introducing any system by which additional restriction would be placed upon the Kaffir, is very debatable. It might have the effect of creating a bad impression in the Kaffir districts concerning the conditions of labour in the Fields, and so operate to retard the inflow of Natives at its source (WCM, 1889, p. 10).[13]

Management was not only interested in avoiding 'a bad impression in the Kaffir districts', but in making a good one. The provision of locations as opposed to closed compounds was one way of doing so. The Buffelsdoorn Gold Mine had one of the earliest family locations. In praise of it, one witness before the ICE said:

> On account of the isolated position of the Buffelsdoorn Gold Mine and the attendant difficulty found in the procuring and retention of an efficient supply of native labour to carry on mining operations, a part of the policy of the company to overcome this impediment has been to always make the surroundings of the natives as comfortable and pleasant as possible. In pursuance of this, five years ago a location ...one mile from the works was established (ICE, p. 376).

Furthermore, as Africans have made their way to the mines of their own will (either that of the labourer or that of the chief), they were under no particular compulsion to endure what they regarded as hardship or abuse. They were at leisure to seek out the most advantageous engagements and remained either until they have made the amount of money they have come for,[14] or until they learnt of better payment or conditions elsewhere. Management also had to contend with the fact that in the same way as many Africans went to the mines of their own will, so a great many more remained in the countryside of their own will.

Against this casual to-ing and fro-ing management eventually, and much to its own distaste, introduced a strategy of making advances to potential labourers. These took the form, mostly, of either cash or cattle, depending on which particular economy the labourer was to be recruited from (Jeeves, A., 1985, p. 5). The intended effect was that, on the one hand, it would entice workers out of the African economies, and, on the other, that it would tie the worker to a particular employer for the period of his contract. The strategy called for a large initial outlay and complex administration and was not necessarily efficacious.

> Although the mine owners knew that cattle advances cost huge sums and led to much abuse by defaulting mineworkers, they found themselves compelled to tolerate the system for some years, so great was their need for labour (ibid.).[15]

The freedom to choose whether to labour or not to labour was not one which capital found itself bound to uphold, especially if it cost it so dearly. But this was not to last, for very soon capital, whose need for labour was far from being satisfied by this trickle, was looking for ways of compelling Africans to labour in the mines.[16] The strategy was about to change from enticement to coercion. But its ability to do so depended upon the consolidation and effective exercise of state power. Major re-alignments and consolidation occurred throughout the region from 1895 to 1902 and further to 1910.

The strategy to compel workers to the mines took two forms: (i) dislodging them from the rural areas; and (ii) keeping them on the Witwatersrand. (i) This strategy had a number of components to it: Already by 1889 some managers were campaigning for a pass system (not to be confused with the later Pass Law of the Witwatersrand) as a form of semi-indenture:

> A regular pass system has many recommendations; By which (1) a free pass, valid for 10 days, would be issued to any Kaffir entering the country, authorising him to look for work; (2) No pass to leave the district would be issued to any Kaffir without an employer's discharge certificate (sic) of recent date; and (3)

Any Kaffir found without a valid pass would be bound for a month to the first employer on the register of applicants (WCM, 1890, p. 9).

The mines did by no means speak 'like one man' on the issue of passes. As already suggested by the passage cited before, public opinion in the 'Kaffir districts' did count for something and some mine owners were sensitive to this (or, at least, their capital was). Others called for a poll tax, 'to ensure that no African remained outside the imperatives of the cash economy' (Legassick, M., 1983, p. 178). Mine managers wanted to make sure that no African was too wealthy to avoid wage-labour. A poll tax was one way of siphoning off such 'excess' wealth (see especially the evidence of George Albu, mine owner, in ICE, 1897, p. 22). According to a Commissioner of the Industrial Commission of Enquiry of 1897, 'The kaffir fee is 10s. per hut, and 2s. 6d. for road tax, and he may take as many wives as he likes, and if he keeps all the women in one house, he has only to pay for one hut' (ibid., p. 129). Land restrictions, although more campaigned for by farmers than by mining capital, did have the effect of undermining the traditional economies and of ejecting their young men as migrant labourers. White rurally-based capital itself, though, existed in at least two forms, and these had contradictory demands on the black population. This put the state in a vacillating position which made it even less effective than it already was.

It was mining capital, ironically, which was to benefit first and most from land restrictions *on an effective scale*. These did not occur in any of the territories which were later to constitute South Africa, but in Mozambique.[17] Portuguese colonial rule in Mozambique, which had always been tenuous, was vigorously strengthened after the defeat of the African chiefdoms in the Luso-Gaza War of 1895-7. In terms of the Transvaal-Mozambique Agreement of 1897, the Portuguese so ruthlessly enforced a policy of land restriction that the South African Republic government was able to guarantee the mines sufficient labour from that source.[18] (ii) Contrary to popular conception, the mines, during this early period, were not interested in a labour force migrating pendulum-fashion between the rural areas and the mines. Harris makes a strong case against this, showing that 'migrant, as opposed to stabilised labour, was not cheap in the nineteenth century' (Harris, op. cit., p. 142). Many mine managers came out against the closed compound system and in favour of locations, i.e., 'normal' urban settlement. Before the Commission of Enquiry of 1897, mine owners, when specifically asked whether they wanted locations as opposed to closed compounds on the Rand, unanimously declared their preference for a permanently settled black workforce on the Witwatersrand. From that Report, we may quote the following evidence:

Commissioner: *With regard to native labour, do you consider that in the cold weather there would be a sufficiently continuous supply unless locations were formed on the Rand?*
George Albu: My experience is that it was never owing to winter that we had less natives than in the summer. There are always other reasons why the natives went home—ploughing, sowing or getting married.
C: Do you consider that if there are locations on the Rand you would have an increased supply of skilled labour?
GA: Yes, I think they would stay here. ...I think if natives had their locations here, and had their wives and families, they would make this place their home (ICE, pp. 23-4);

Commissioner: *What do you think of the compound system that prevails in Kimberley?*
George Albu: I would not recommend the compound system.
C: *Why?*
GA: Because I think it would hurt the commercial industry (ibid., p. 25);

Commissioner: *Can you suggest any plan by which a permanent supply can be relied upon for the Rand, skilled principally?*
Edward Hay: The only way is to give natives facilities for family life. We do it to a certain extent on the George Goch [Mine], and we get into considerable trouble for doing it. We have a location upon our lower claims, and I have boys who have their wives and families, who have been working for the mine for the last eight years. If locations could be established somewhere in the neighbourhood of the mines — within walking distance — so that the natives could bring down their wives and families, I think you would have a far greater supply than you require (ibid., p. 43);

Commissioner: *You have a lot of experience of the compound system in Kimberley?*
Sydney Jennings: Yes.

C: *Would you think it would be advisable to apply the same thing here?*
SJ: Taking all considerations, the commercial community and the good of the land, I would say no. We could get sufficient control over the kaffirs if the Pass Law was efficiently administered, without the compound system.
C: *And how about establishing kaffir locations along the Rand?*
SJ: I would say it would be a most excellent idea if the sale of liqour in the locations was absolutely prohibited... (ibid., p. 46);[19]

Amandus Brakhan: ...There are many ways in which the Government can assist the mining industry in its endeavours to procure a sufficient and cheap native labour supply. I will only mention the encouragement of locations not far from the Rand... (ibid., p. 184);

George Denny: We also think that the introduction of a law establishing large native locations in the vacinity of the mining camps would be a great advantage in keeping up the permanent supply of labour. ...There seems good ground for assuming that if locations were established in the proximity of the mines, there would be a base formed for a working force more permanent and steady in character than exists at present (ibid., p. 376).

One witness even went so far as to describe migratory labour and compound living as an 'unnatural condition' (William Hall's evidence, ibid., p. 429). This does not mean that compounds did not exist on the Rand at this time. The settlement pattern of the mine workforce was that most white miners, being either single or without their wives, lived in boarding houses or company 'single quarters', while black mine workers, virtually all on short-term contracts and by themselves, lived in compounds on mine property.[20] The pre-war compounds had no particular significance in respect of political economy, being merely a housing type. Their occupants were free to come and go as they pleased.

The closed compound system was introduced in Kimberley primarily to counter diamond thefts. Thefts of gold amalgam (an intermediate product) did take place on the Rand. One estimate puts the value of such theft at £750,000, i.e., 10% of the annual output (ICE, p. 456). It was hence sufficiently serious a problem for the mines to want something done about it. But their proposals were always confined to the areas of law and law enforcement.

They wanted a revamped Gold Law and better detectives. Never has it been suggested that tighter control on the compounds might reduce the amalgam thefts, for the culprits were not the black workers, most of whom were concentrated underground, but those in the plant on the surface. In one mine owner's opinion, 'it is done ...by the amalgamators, because the amalgamators are the only ones who could handle the amalgam' (Joseph Robinson's evidence, ibid., p. 263). We would therefore suggest that Richardson and Van Helten might have been overhasty in claiming a direct connection between the increase in amalgam thefts and the reduction of black wages in 1896-7 (Richardson, P. and Van Helten, J-J., 1982, p. 92). Indeed, one mine owner, in an exchange with a Commissioner at the ICE, explicitly denies a connection between black workers and amalgam thefts:

> If you thrust on individuals or companies the responsibilities and duties which are generally supposed to devolve to Government [detection of and prosecution for gold amalgam thefts —FS], you would give the directors of the mines the right to institute the compound system, to keep the boys on the property, and the searching system, so that amalgam thefts would be detected; but those are systems that we would never suggest (James FitzPatrick's evidence, ICE, p. 63).

We believe that the issue of gold thefts can safely be dismissed as having played any role in the evolution of control over the black workforce on the mines. According to Van Onselen, 'a relatively relaxed compound system emerged on the Rand in the years before the South African War' (Van Onselen, C., 1976, p. 131). The mine-owners' concern that confining the workers to closed compounds would harm the 'commercial community' suggests that the workers were free to spend their money as they wished. If Williams is to be followed, then this freedom to engage in exchange should, at least, have given rise to a conflict, if not actually expressed a contradiction. For Williams, as the variable capital emerges from production already realised, 'they can just as easily hold onto this money and purchase the means of subsistence for the workers. As the producer of the money-material there is no need for the worker to constitute an integral part of the circulation process' (Williams, op. cit., p. 24). As will be shown below, such a contradiction does, indeed, emerge. But this is not addressed in Williams' theory.

Hobson, too, holds a contrary view and asserts that if African mine workers were to live in the advocated locations, then, 'the houses they will occupy will be the property of the mines, as also the shops where they will be compelled to deal. This has been the policy advocated by the chief mining experts' (Hobson, op. cit., p. 271). As has already been abundantly shown, it is true that the mine owners advocated locations, but Hobson provides no evidence, neither could we find any, to suggests that the mines conceived of their own black workforce as potential captive consumers of everyday commodities. Indeed, quite the contrary. The mines were forever mindful of the interests of the 'commercial community'. They advocated that *the state* provide these locations. Most of the literature, Van Onselen being the notable exception, do not appreciate the evolution of the compound system and the gradations that developed within it. Thus does Williams, e.g., talk of black mine workers, in both diamond digging and goldmining, as being 'herded into compounds' right from the start (Williams, op. cit., p. 7).

Capital saw urban settlement as a strategy to counter the labourers' tendency to return to the countryside. Although proletarianisation was clearly underway, the form of labour can at best only be described as transitional. Right from the outset, since each capital had to individually induce its own labourers one by one out of another economy (and at no small expense), it had no interest in losing that labour to a general pool of 'free' labourers. Particular labourers were engaged by particular capitals for particular jobs. Those who did leave one employer for another were guilty of 'desertion', and desertion was one of the biggest problems for goldmining capital. A market for labour did not exist, and as we shall see, could not be allowed to exist.

Africans did respond to the presence of market economies in close proximity, but related to them in such a way that the market economies remained ancillary to their own traditional economy. The peculiar role that the countryside, or, more particularly, the Reserves came to play in the economics of goldmining only suggested itself later, as capital struggled to control the value of the labour-power of the workforce it was as yet only trying to create.

Two environments and the question of free wage-labour

There were two peculiarities about the price of gold in the late nineteenth century: the first is that the price was fixed; and the second is that it was fixed at a level which bore no relation to its value.[21] The first of these was important in our discussion of competition in goldmining. But it is the second to which we now turn in order to explore the significance of the *value* of goldmining labour-power for gold production in South Africa.

The literature proposes a link between the fixed gold price and the level of black wages on the mines. Mostly, the argument runs as follows: The fixed gold price meant that the high cost of gold production could not be passed onto the consumer. The mines, therefore, necessarily had to strain every sinew towards cost reductions. This was more effective in some areas than in others. The area where cost reductions could be most effectively pursued was in the area of wages, both black and white. White workers, however, either through their skills or through an alliance with the state/capital, managed to escape most of this. Black labour, consequently, had to bear this burden. Their wages were thus extremely low when compared to white wages, were the first to be attacked, and, underwent virtually no real increase from the eighteen nineties until the nineteen seventies.

If labour-power is accepted to be a commodity, then it must be discussed as a commodity. What all the literature lacks is a consistent treatment of labour-power as a commodity. It is treated as a commodity only in so far as it has a price and the precise magnitude of that price is subject to the laws of supply and demand. The story of black labour on the gold mines thus reduces itself to one of a struggle between capital and labour over the *level* of African wages. Duncan Innes, e.g., says that the high labour costs, 'could not be resolved without restructuring the labour-supply in such a way that either the supply increased or the demand decreased. These were the only ways in which wages could be reduced...' (Innes, op. cit., pp. 28-9).[22] Innes appears to consider the price of labour-power to be secular. The question of value never arises for him. Neither does it for Peter Richardson, whose discussion of the importation to the Witwatersrand of indentured Chinese labour is examined below.

Although not equipped with the term 'value of labour-power', the category lay at the centre of goldmining capital's conceptualisation of the wages problem. In a conceptualisation which departs from exchange value as the fundamental category, the value of labour-power can only be expressed as the cost of living.[23] The incisiveness of capital's dealing with this question comes across most pristinely in the statement by William Hall, California civil engineer, to the ICE of 1897. Hall explains that:

> The cost of living must be a fair index of the wage to be paid, in order that the workman remain contented in whatever capacity he may be serving. Those who labour only with their hands receive but small margins over living rates, while those who have some specially valuable professional or managerial ability demand and get wide margins. Nevertheless, living expenses is an index to the wage or salary in each grade of service, separately, and for each locality (William Hall's evidence, ICE, p. 420).

This spokesperson of capital is a theoretical giant against Williams and Innes. Not only does he capture the concept of the value of labour-power and lay it at the heart of the wages question, he also appreciates that the value of labour-power is different for different categories of labour, and that the value of labour-power varies geographically. This very point is given careful theoretical examination by the liberal economist Edna Bonacich in her article *A Theory of Ethnic Antagonism: The Split Labor Market*, which appeared in the journal *American Sociological Review* (Bonacich, op. cit.). Bonacich's starting point is a simple one: 'To be split, a labor market must contain at least two groups of workers whose price of labor differs for the same work, or would differ if they did the same work' (ibid., p. 549). She deals with the value of labour-power as the 'price of labour', which refers to:

> labor's total cost to the employer, including not only wages, but the cost of recruitment, transportation, room and board, education, health care, ...and the cost of labor unrest. The degree of worker 'freedom' does not interfere with this calculus; the cost of a slave can be estimated in the same monetary units as that of a wage earner, from his purchase price, living expenses, policing requirements, and so on (ibid.).

At this point in history, the reduction of black wages was not only consistent with capitalist economy, it was also consistent with economy in general. This economy was expending more than the necessary portion of its aggregate human labour on the reproduction of this particular kind of labour-power. Its reduction *in that place, at that time* was a progressive step; an anti-waste measure.

Innes and so many others see the question only within the narrow confines of trade union wage struggles. Other developments are then explained in terms of this struggle. Thus, e.g., are the

Labour Reserves conceived as enabling capital to reduce the level of wages, and the compounds conceived as preventing their occupants from organising to increase it. The literature appears to be more concerned with the division of surplus-value than with its production.

If we are concerned with surplus-*value* production, then we must be concerned with the *value* of the labour-power from which springs that surplus. Capital is interested in surplus-value production, and early South African goldmining capital was certainly most interested in the value of labour-power. With this known, capital would know the capacity of that labour-power to yield surplus-value. It was a simple matter of a commodity's technical specification. The structural shortage of unskilled black labour on the Witwatersrand right up to the Act of Union in 1910,[24] together with the fact that most Africans brought their *surplus* labour-time, rather than their *necessary* labour-time into the circuit of capital, meant that the value of their labour-power could not, by the very nature of the case, be known. Skilled white labour was in short supply too, but white labourers brought their necessary labour-time into the circuit of capital. Computing the value of their labour-power was therefore a straightforward matter.

For Wolpe, though, the crucial distinction is that white workers have no access to the means of production, while black workers do. These means of production, for Wolpe, exist in the Reserves. The environment within which the white worker reproduces himself is restricted to the urban environment, while his black counterpart does so in an area comprising both the urban environment and the Reserves. This enables capital to pay the worker less than the value of his labour-power on the assumption that the difference is made up for by Reserve production. Wolpe puts it thus:

> When the migrant labourer has access to means of subsistence, outside of the capitalist sector, as he does in South Africa, then the relationship between wages and the cost of production and reproduction of labour-power is changed. That is to say, capital is able to pay the worker below the cost of his reproduction. In the first place, since in determining the level of wages necessary for the subsistence of the migrant worker and his family, account is taken of the fact that the family is supported, to some extent, from the product of agricultural production in the

Reserves, it becomes possible to fix wages at the level of subsistence of the individual worker (Wolpe, op. cit., p. 434).

One problem with Wolpe is his lack of periodisation. Despite the 'dominance' of the capitalist mode of production over the pre-capitalist modes, this dominance appears to be static. After all, up until the discovery of diamonds, capital had but a toe-hold in southern Africa. From the discovery of diamonds till the eve of the unification of South Africa, although capital was on the ascent and the African economies steadily being undermined, the 'dominance' of capital was not yet a clear-cut matter. But the relationship then was different from what it was prior to the mid-1870s. Today the Reserves have all but collapsed. What is there for the capitalist mode of production to dominate? Part of Wolpe's problem is that he does not say what this 'dominance' consists in. He does not say what exactly it is about the capitalist mode of production which makes it 'dominant' over other modes of production. Another part relates to his object. He tries to give an account of the labour Reserves without rigorously examining and applying the category value of labour-power. In other words, for Wolpe, the value of labour-power must serve to explain the Reserves, rather than the Reserves giving expression to certain developments hinging around the value of labour-power.

We shall restrict our response merely to making a few brief points. Much theoretical headaches have been suffered over how to account for workers who have access to the means of production. This conundrum has its origin in a notion which has taken root in a particularly mechanical brand of Marxism, that the proletariat, by definition, is 'freed' from the means of production. The point is not whether he has access to the means of production, but whether his *necessary* labour expenditure sets those means of production in motion *inside or outside of the circuit of capital*. If this is done within the circuit, then the worker could own the entire factory, he would still only be able to reproduce his labour-power when summoned to do so by capital, which owns his labour-power. Where the use of such means of production occurs outside the circuit of capital, then the worker can only be allowed to exercise his *surplus* labour-time on it. In the case of the South African labour Reserves (which is not within the concerns of this thesis) it is the entire labour-time of the worker's wife and children, plus a portion of his own *surplus* labour-time which are brought to bear on the means of production in land.

It is more usual for workers to have access to some means of production than not at all, be it a plot of land in the case of the Third World, or a box of tools in the case of the industrialised world. Telephones, sewing machines and cars — all owned by workers — are all means of production. Access to by the workers to their own means of production is on capital's terms. It is capital's fine-tuning of the working class to its requirements. Workers with 'access' to the means of production under conditions of capitalist production find that such access does not free them from capital. It is, indeed, capital which has such access to these means of production *through the worker*, the worker serving merely as the most cost-effective caretaker of them. A great deal of mysticism surrounds this question and many others relating to the Reserves. But this is not the place for trying to examine these.

What is interesting about Wolpe's conception is not so much that he sees (migrant) black workers and white workers as belonging to different reproductive environments (a number of writers do that), but that for blacks the *relationship* between wages and the value of labour-power is different from that for whites. Wolpe's general theory may be an unconvincing one, but his starting point is correct. This is the same as saying that black labour-power and white labour-power each has a different relation to capital. The same is said by Richardson and Van Helten, and the same is said by Ticktin.

Ticktin's real exploration of the category of value of labour-power is when he makes a distinction between the value of labour-power in the colonies and the value of labour-power in the metropolitan areas. For Ticktin,

> There is a particular problem when it comes to the wages of the workers in the colonies. They receive lower wages than their metropolitan counterparts, permitting a higher rate of surplus value even if the local agents may receive only a small proportion of that surplus value. On the other hand, the value of their labour-power is nationally determined, not internationally, so that they may actually not be superexploited, though the rate of extraction of surplus value may be very high (Ticktin, op. cit., p. 74).

It must first be clarified that by superexploitation, Ticktin means, 'paid below the value of their labour power' (Ticktin, op. cit., p. 76). The proposition is sound and more coherent than in the case of either Wolpe or Williams, but it cannot simply be put in this form if we are to examine its working-out in a particular context. Three such contexts are identified and briefly discussed by Ticktin: the superexploitation of minority workforces, sometimes immigrant, sometimes indigenous, in the metropolitan countries (Moroccans in France or blacks in the USA); 'the colonial worker'; and the superexploitation of the majority workforce in 'a "settler economy", such as contemporary South Africa' (ibid., pp. 76-7). For Ticktin, the payment for labour-power *at* its value is a feature only of industrial capitalism. He says that, 'In the non-industrial stage of extraction of surplus value the metropolitan power had no interest in preserving the lives of the workers who were then often paid below value' (ibid., p. 75). But the proposition itself provides the important qualification that, 'the value of their labour-power is nationally determined, not internationally, so that they may actually not be superexploited'.

Ticktin could have made a more forceful argument by enforcing his qualification and assessing exploitation against the background of a 'nationally determined' value of labour-power, in order to determine whether such exploitation ought to be characterised as superexploitation. His argument, however, proceeds to assume superexploitation in all instances. A strong case can be made that, 'In the non-industrial stage of extraction of surplus value the metropolitan power had no interest in preserving the lives of the workers who were then often paid below value', but then 'non-industrial stage of extraction' needs first to be differentiated into a *pre*-industrial stage of extraction, and a *post*-industrial stage of extraction, i.e., under the dominance of finance capital.[25] The most important differentiating factor being that in the former, one is dealing with a world of relatively atomised production, while in the latter, the production which lends the era its character is global and integrated to a very high degree. In a pre-industrial stage of extraction, the worker is a particular unit with a particular skill in a particular place going at a particular price, i.e., an individual of the particular; in the post-industrial stage, both capital and labour move around the globe (of course capital with much greater facility than labour) with labour-processes so standardised that virtually any workers anywhere can take almost any employment from any firm. In both instances is capital able to take advantage of the difference between 'nationally determined' values of labour-power — *but so, too, are the workers able.*

It could be argued that the systematic undermining of a society's means of production amounts, in an overall sense, to payment below the value of labour-power, since incorporated into the value of any labour-power is the objectified labour of that society's prior history. The worker does not only have to reproduce himself and his family, but also his entire society, i.e., the means of reproducing his labour-power in its current condition. Just because a worker and his family physically survive from generation to generation, does not mean that he is not being paid below the value of his labour-power. This could be one possible argument in support of the superexploitation thesis. In an urban-industrial environment, one would have to show decline in the worker' standard of living. A Moroccan worker living in squalor in France might just as easily have been living in squalor in Morocco. Perhaps his condition in France is worse than that in Morocco, but perhaps this worker is sending part of his wages to his family in Morocco, who through this are able to extricate themselves from Moroccan squalor. Likewise Phillipinos on the high seas, or Pakistanis in Kuwait, or Zimbabweans in South Africa. Without figures, these relations cannot become clear.

Only Trewhela relates the value of labour-power of the gold miners directly to the nature of gold as money-commodity. By interfering directly in the reproduction process of goldmining labour, says Trewhela,

> Gold-producing capital infringes even on the value-character of labour-power itself. The market thus appears as shrunken and restricted at both poles of the process of gold-production simultaneously. As product, and as producer, at the end of the production process and at its beginning, basic elements in the life-cycle of gold-producing capital are reproduced in these non-market forms (Trewhela, P., 1986a, p. 17).

But this 'infringement' on the value-character of labour-power is far deeper and broader than Trewhela appears to appreciate. It constitutes a *form* in itself and relates directly to what Trewhela elsewhere very accurately refers to as 'a bizarre kind of socialism of value' (ibid., p. 16). The existence of the labour Reserves, from which Trewhela makes this correct, but nevertheless intuitive insight, are but the circumstantial expression of something much more basic. The present essay shows that this 'bizarre kind of socialism of value' is far more fundamental than the mere geographical organisation of exploitation. It is already present *as a potential* in gold's being 'both a commodity and not a commodity', i.e., both a value and not a value; both socialism and not socialism.

Without a study of the relationship between particular use-value and universal use-value, and, especially an appreciation of the distinction between them, there is no fixed link to bridge the gap across which Trewhela so brilliantly leaps. The spanning structure is the nature of capital which realises itself through the production of a commodity for which the contradiction between use-value and exchange value has been solved. Goldmining labour is not special on account of gold being special, but on account of goldmining *capital* being special. This, too, Trewhela does indicate, but again, only intuitively:

> the entire system of social relationships in the African subcontinent, ... with the gold mines at the centre, rests upon the contradiction of capital's promotion of the market and its infraction of that market through the migrant labour system the pass laws... [etc.] ... (ibid., p. 18).

The relationship of wage-labour is with capital, and not with the particular commodity that capital sets it to. This is the *differentia specifica* of wage-labour. By considering only gold's universal use-value and ignoring its particular use-value ('I exclude from consideration here the use of gold for non-money purposes, e.g., dentistry') (ibid., p. 15n),[26] the character of commodity-producing labour as both concrete and abstract labour slips from view. All that remains is the character of the money-commodity as immediately social. The labour which produces it is then taken to be immediately social. It is forgotten that such labour produces not the money-commodity, but precisely those commodities which have been 'exclude[d] from consideration', and without which it could never be the money-commodity. Williams' treatment of the gold and labour question makes it even more obscure than its appearance causes it to be. To build a thesis on Williams, as Trewhela does, is to attempt to turn alchemy into chemistry.

Wolpe's words, 'since in determining the level of wages necessary for the subsistence of the migrant worker and his family, account is taken of the fact that the family is supported, to some extent, from the product of agricultural production in the Reserves, it becomes possible to fix wages at the level of subsistence of the individual worker' (loc. cit.), are more instructive than he appears to realise. He does not ask himself how or when capital went about, 'determining the level of wages necessary for the subsistence of the migrant worker and his family', i.e., determining the value of the migrant worker's labour-power.

Capital has every interest in determining the value of labour-power as accurately as possible. Let us assume a situation where capital is under pressure to reduce its cost of production and the most effective area for doing so is in wage reductions. This it will normally establish in the market soon enough, when it finds that it cannot secure the commodity (labour-power) that it needs for the value that it is prepared to cede. As in all exchange, the threshold between socially necessary labour-time and surplus labour-time is always tested by both parties to the act. Bonacich goes straight to the economic question: 'The prejudices of business do not determine the price of labor, darker skinned or culturally different persons being paid less because of them. Rather, business tries to pay as little as possible for labor, regardless of ethnicity, and is held in check by the resources and motives of labor groups' (Bonacich, op. cit., p. 553).[27] If, on the one hand, the level of wages is too high to reflect the value of labour-power (as they were for black wages under conditions of such acute shortage), then capital would want to know the extent to which, under normal supply conditions, it would be able to reduce these wages. If, on the other hand, wages were being paid at the value of labour-power, it knows that in order to reduce the level of wages, the value of labour-power itself would need to be reduced. The *value* of labour-power, as distinct from its *market price*, would be one of the first things that capital would need to ascertain.

None of the works assessed for this study make this point. The attempts of goldmining capital to compute the value of labour-power down to exact pounds, shillings and pence seems, at first, to border on miserly obsession. A full account of this may be gleaned from the ICE (*passim*). The evidence of James Hay to the Commission puts the problem succinctly:

> **James Hay:** ...In working those mines ...wages are also divided into two classes, whites' wages and blacks' [wages]. The cost of white wages depends upon the law of supply and demand, and the cost of living of the white people. ... The cost of living of the white people depends upon the price of food they consume, and that is made up in this country largely by the charges on railway carriage and Customs dues, in addition to the original cost of the goods at the place where they came from. The next question is that of native wages, and it is a very important one, not only for the mines, but for the farmers, and everybody who

requires labour. I think we are all agreed that the natives are too highly paid in this country. But the difficulty is, how can the wages by reduced… (ibid., p. 4).

The value of white labour-power is known, or at least can be known; the value of black labour-power, on the other hand, is not only unknown, but also cannot be known. White workers brought their *necessary* labour to the mines. The limits of that necessity are therefore testable by capital. It knows both how and to what extent white wages can be reduced. Not only does the boundary between necessary and surplus labour for blacks lie outside of the circuit of capital and hence beyond capital's ability to test, but capital is unable even to approach this on account of supply and demand, i.e., there still being too much competition for black labour.

The truth of this was brought home to capital with its first attempt at reducing black wages. Harris, in analysing the relationship between the African economies and the mines, gives the following account:

The benefits of the redistributive economy …provided them with the land and labour needed to subsist outside the wage economy. This acted as …a weapon with which to combat low wages and poor conditions on the labour market. Thus when the Chamber of Mines tried to reduce African wages in 1890-91, thousands of Moçambicans were able to withdraw their labour by returning home and in this way were able successfully to force a return to the old wage level (Harris, op. cit., p. 157).

This clearly does not stem simply from black miners being in great demand. Workers can be in great demand and still be unable to withdraw their labour, simply because they would starve. This was not the case here. They were fully able to reproduce their labour-power with or without mine labour. Their necessary labour was exercised outside the capitalist economy. Or, as Bonacich puts it, 'The worker's standard of living does not, therefore, depend on his earnings on the job in question, since his central source of employment or income lies elsewhere' (Bonacich, op. cit., p. 551).[28] All labour they exercise outside of their own economies was therefore superfluous to their needs (until such time, of course, as their needs were to expand or their economies to shrink). Therefore all monies they earned were superfluous to their needs. The mine owners took the trouble to study this question. One of their number, giving testimony before the ICE, puts it thus:

> **George Albu**: The reduction of native wages is necessary for two reasons, the one is to reduce our whole expenditure, and the second has a very far-reaching effect upon the conditions which may prevail with regard to native labour in the future. The native at the present moment receives a wage which is far in excess of the exigencies of his existence. The native earns between 50s. and 60s. per month, and then he pays nothing for food and lodging, in fact, he can save almost the whole amount he receives. At the present rate of wages the native will be enabled to save a lot of money in a couple of years. If the native can earn £20 a year, it is almost sufficient for him to go home and live off the fat of his land. In five or six years' time the native population will have saved enough money to make it unnecessary for them to work any more. The consequences of this will be most disastrous for the industry and the state. This question applies to any class of labour, and in any country, whether it be Africa, Europe or America. I think if the native gets sufficient pay to save £5 a year, that sum is quite enough for his requirements, and will prevent natives from becoming rich in a short space of time.
> **Commissioner**: *You say the native does not require luxuries, and if he has worked for a year he has saved enough to go back to his kraal and remain idle?*
> **GA**: Yes.
> **C**: *Can you suggest any remedy for this?*
> **GA**: The only remedy I can suggest is that we pay the native a wage which, whilst enabling him to save money, will hinder him becoming exceptionally rich.[29]

To most scholars this would stand as an example of extreme callousness on the part of goldmining capital. The truth is that capitalism did provide the African societies with an opportunity to convert a great deal of their *surplus* labour into wealth. To be wealthy in a society with relatively few needs to address, is not the same as being wealthy in a society with relatively many needs. Capital is never interested in paying more than the value for labour-power (or for any other commodity, for that matter). In this case it was definitely paying for labour-power higher than its value. It was concerned to bring the price it paid for this commodity more in line with its value.

The wage is the amount of value, in money terms, which expresses the equivalent of the worker's necessary labour-time. The only way in which capital can know the quantity of that necessary labour-time is by bringing its exercise *into the circulation process* of capital. How was this to be done? To borrow Hobson's words,

> There are only two genuinely economic forces which will bring such labour more largely into the labour market: The growth of population with increased difficulty in getting a full easy subsistence from the soil is one; the pressure of new needs and a rising standard of consumption is the other (Hobson, op. cit., p. 255).

'Genuinely economic forces' are things which take time to run their course and capital is not something which can always wait. If the population is not growing fast enough to exceed the capacity of the land to support it, then the land must be reduced till it can no longer support the population; if new needs are not being added fast enough to exhaust the social product, then that social product must be reduced to the point where it can no longer support the needs. Put differently: either destroy the capacity of the African economies to engage all of the worker's necessary labour-time; or seize the labourer's surplus labour without exchange, thereby leaving him with the product of only his necessary labour.

The systematic emasculation of the African areas took place through land dispossessions and the extraction of a rent. The latter took two forms: 'taxation' and price fixing. The various taxes imposed upon different African economies (hut tax, dog tax, poll tax, etc.) were not taxes as ordinarily understood. They had little to do with state revenue and no state services were had for them in exchange. When revenue was mentioned in this context, it was purely incidental:

> According to the native laws the actual tax per native in the native districts amounts to about £2 10s.; the only exception to the liability of this payment being natives residing as servants under white persons. Thus, if the dignity of labour were impressed upon them by the enforcement of this law, we are likely to get a larger supply. The other purpose served would be — that such boy coming to work would earn money, and after completing his term of service he would have sufficient funds to enable him on his return to his district, when called upon by the

Government, to pay all his taxes, which would be an additional revenue to the Government. (Charles Goldmann's evidence, ICE, p. 117).

The object was to siphon off the surplus labour of the African economies. Or, as Hobson puts it, 'not to provide revenue, but to compel labour' (Hobson, op. cit., p. 268).[30] But it was no use this surplus labour being offered to the state as a *corvée* (the state's own labour needs notwithstanding).[31] Neither was it efficacious for this labour to be seized as particular commodities, e.g., cattle or maize (the shortage of these, too, notwithstanding). Surplus labour had to be handed over to the state in the form of money, hard cash. Only two opportunities existed for Africans to transform surplus labour into money. Either the labour was objectified into agricultural produce and disposed of for cash in the market, or living labour itself was sold for a wage. The existence of the first possibility made the second less effective. 'All ...engaged in the collection of native taxes, should discourage the system of the natives selling their cattle to meet the tax, and by their paying only in coin, there would be a greater necessity for them to come out and work', urged the mine owner George Denny (George Denny's evidence, ICE, p. 375). Even those African economies which still traded their surplus produce by barter, quickly switched to cash. The problem was solved both by further undermining the productive capacity of the land through increased dispossessions and by through fixing of the price and centralised buying of African produce. This amounted to the extraction of a rent by settler agriculture from African agriculture.

For those who went into wage-labour, the handing-over of part of their earnings to the state amounted to a neutralisation of at least part of their surplus labour. A portion of their labour-time was removed from the surplus side of their economic balance sheets and added to the necessary side, thereby shifting the boundary between surplus labour-time and necessary labour-time in favour of the latter. Part of the worker's necessary labour-time, though not yet all of it, was now spent *inside* the circuit of capital. And since this was the object, the 'tax' in fact became superfluous once the worker becomes established as a wage-labourer. Mine owners could see this, as George Albu, e.g., insisted before the 1897 Commission of Enquiry:

Commissioner: *If a man can live without work, how can you force him to work?*

George Albu: Tax him then. If I have £5 per month to spend, I don't want to do any work; but if the Government passes a law that all gentlemen at large... must pay £3 per month tax, there only remains £2, and I am forced to work (ibid., p. 22).

A hut tax of £2 was imposed in the ZAR in 1895 (Hobson, op. cit., p. 269), but by 1897 a Commissioner on the ICE found it necessary to object to yet another call for taxation of blacks by pointing out that, 'They are already taxed higher than the white population, and it would be unreasonable to tax them much more than they are taxed' (ICE, p. 230).

But these efforts at coercing labour out of the African economies began to prove successful. A sufficiently large influx of Africans to the Rand was underway between 1895 and 1897 for the mines to declare, 'at the present moment we have sufficient labour' (George Albu's evidence, ibid., p. 14), and for the phenomenon to be described as a 'Kaffir Boom'. But this influx did not bring the thriving location settlements which the mines were campaigning for. This was a time when releasing workers from the Rand, or, indeed, from the premises, was without guarantee that they would return. Even by 1889 the Chamber stressed that:

> the local mining industry is subject to outside competition in the Native labour market. Other mining districts offer wages almost as high as the Rand, the Railways being constructed in Natal and the Cape Colony draw away large number of Kaffirs, and the Boksburg Tramway is employing 1,500 to 2,000 Kaffirs at very high rates of pay (WCM, 1889, p. 10).

By 1904, 'three-quarters of a million peasants in the Transvaal had access to land on terms more favourable than the 123,000 on Government locations or the 50,000 Africans in full-time employment' (Bundy, C., 1979, p. 209). Between the end of the South African War and the imposition of the Native Labour Regulation Act of 1911, about 25,000 Africans in the Transvaal went to seek *temporary* employment each year (ibid., p. 212). Although part of this would be accounted for by agriculture, rail and road-building, the largest employer at the time was, nevertheless, the gold mines on the Witwatersrand.

Such workers, emerging as they did out of a Spartan environment, were now immersed in an ocean of new wants. The contradiction now is that the condition which allows for the value of labour-power to be known is the very condition which makes that value fluid.³² And in the specific condition of the black workers with their relatively frugal needs, that fluidity meant increase. These new needs are then assimilated into their standard of living. Thus does the value of their labour-power increase. Of course, when they acquire new needs they are also prepared to work more to meet those needs. The point is made by Henry Jennings, mine owner, when recounting his experience with locations in Venezuela,

> They came from different islands, and they had these little locations. There was a certain amount of rivalry between them, their wants increased, and that stimulated their ambition, and they were finally willing to do more work than at first (ICE, p. 229).

But such increase in the value of labour-power, higher absolute surplus-value notwithstanding, runs directly against the solution to the second labour problem — that of the reduction of costs. The problem was exacerbated by the fact that from about 1895, the gold mines no longer offered the only means of making a living on the Rand. There were increased opportunities for both employment and self-employment, small as these were in comparison to mine work. Charles van Onselen describes the situation of the 'houseboys' (general domestic workers) and the self-employed washermen in the 1890s around the time of the so-called 'Kaffir Boom'. Houseboys were earning on average a monthly wage of eighty shillings in contrast to the average fifty shillings per month of black mine workers. Although van Onselen does not give a figure for washermen's income, it is clear that, despite its changing fortunes, hand-laundering did offer a viable occupation. The accommodation of so many white mine workers in boarding houses and the absence of so many European men's wives meant that the market for this particular service was very large. Indeed, large enough for Johannesburg to boast half a dozen steam laundries by 1898 (Van Onselen, C., 1982a, pp. 18-19).³³

The point is that despite the fact that the manner in which black mine labour was recruited militated against the development of free wage-labour, free wage-labour was nevertheless developing around the mines on the Witwatersrand and directly ancillary to it. Normal urban residence for black miners meant their induction into a pool of fluid wage-labour. There is every reason to believe that in such an environment, with the mines under pressure to keep labour costs down, they would quickly have lost their workforce. Somehow, they had to be near to the mines, yet not form part of a general free labour force.

The solution lay in the closed compound system. Although mine owners were as late as 1897 still expressing themselves against compounds for fear of scaring off potential recruits and because of reluctance to antagonise the 'commercial community', the issue was brought to a head by the South African War and its immediate aftermath, when the gold mines found themselves in a severe labour crisis. Wage reductions were pushed through during the South African War while the mines were, in any case, on a much-reduced staff as production had been interrupted. There was very little reason for idle black mine workers with a secure base in the countryside (which most of them had) to remain in a very dangerous war zone. The return of the Rand to production in 1901 found the area in the grip of a serious shortage of labour. For most scholars, the significance of this state of affairs lies only in that it prompted the introduction of indentured Chinese labour in 1904, which served to further exacerbate the race and labour question. This, in our view, misses the most important points: firstly, the significance of *indentured* labour as a form; and, secondly, the role of this labour-form in effecting the changeover from one form of proletarianisation to another.

The problem of creating a world proletariat had its particular expression in South Africa in two areas: sugar production in Natal, and gold production on the Witwatersrand. The specific examination of early Natal sugar production is not of interest here. What is important is that the form of labour was indenture. On a world scale, by the time of the opening of the goldfields on the Witwatersrand, not only was slavery long abolished, but indentured labour was past its zenith around the 1850s.[34] Although a labour force did not exist in the goldmining area and had to be created from scratch, such labour force was never conceived of as either slaves or indentured labour, but as wage-labourers. The peculiarities of lode production under inappropriate gold-price conditions, together with the manner in which proletarianisation was proceeding, meant that the narrow value-specifications required by goldmining capital could for decades not be met. This could be accommodated up to a point, but the upward pressure on the value of goldmining labour-power became intolerable for capital in the wake of the South African War.

The source most cited on the Chinese labour question in early South African goldmining is Peter Richardson's article *The Recruiting of Chinese Indentured Labour for the South African Gold*

Mines, 1903-1908 (1977). Richardson provides a well-researched account of this comparatively obscure chapter in Witwatersrand goldmining history. It is an account, though, which does not attempt to offer a political economy of indentured labour. Indeed, the fact that the labour was indentured is, for Richardson, purely incidental (the word 'indentured' appears only twice in the text). For Richardson, as for scholarship in general, the importance of this labour is that it was Chinese.

To Richardson's credit, though, it has to be said that he does situate this episode in the context of a colonially initiated, 'international movement of labour' (ibid., p. 85). That this particular system of international movement of labour takes over from another which had just come to a close, viz. the slave trade, and that its form stands mid-way between slavery and wage-labour, is not drawn out by Richardson. An institution with the capacity to organise the procurement, transfer and application of labour on a mass scale across the globe is not examined for its theoretical significance. This is despite the very detailed description of the 'high degree of vertical integration' of the company (in fact a central organ of the Witwatersrand goldmining industry as a whole) which, 'operated as recruiting and shipping agency in China, receiving agent in Natal and co-ordinating and advising agent in the Transvaal' (ibid., p. 108).

That indentured labour is less free than proletarian labour is appreciated by Richardson, but again, the significance of this is confined to accounting for the devices employed by the gold mines to conceal that fact. 'The necessity of securing indentured as opposed to "free" labour for the Transvaal mines, had meant the employment of reputable western firms to give the whole operation a semblance of bona fides' (ibid., p. 97). Evidence for the importance of this distinction in political economy is provided by Richardson himself, when he quotes an account of Japanese recruitment of free wage-labour from the same area in China:

> Already difficulties are experienced whenever the Japanese require coolies and send a ship to Chefoo to collect them. All able-bodied men are taken, they are paid good wages, there is no medical examination, no questions are asked and coolies are not tied down by strict regulations. They ...can return when they please (quoted in ibid., p. 106).

This split in the Chinese labour force in the recruiting area of China into a coerced labour force and a free labour force, reflected exactly the kind of transformation taking place in the black workforce on the Witwatersrand at the time. One section of the labour force had the value of its labour-power held fast, while that of the other was free to adapt. Those blacks on the Witwatersrand outside of the employ of the gold mines found themselves the envy of those in goldmining. The same 'difficulties' existed between the two branches of the newly bifurcating workforce in China.

On the Witwatersrand, 1903 saw wage-increases, endemic 'desertions' and an acute labour shortage, and this amidst increased pressure to expand production.[35] The possibility that the mines would ever secure for themselves the stable, cheap labour force upon which their very operations depended was now seriously in question. In 1903, 'most South African blacks persistently rejected mine employment, especially underground, where any alternatives were open to them. Africans knew well enough about conditions on the Rand...' (Jeeves, op. cit., p. 11). In this year and in pursuit of a solution, the goldmining industry, firstly, formed a Committee of Agents to look into the feasibility of Chinese labour importation,[36] and secondly, turned to the diamond industry for help. A special Commission was dispatched to study and report on the closed compound system operated at De Beers. This system served as a model for the closed compounds in which the Chinese workers who started arriving a year later were housed. However, once the Chinese had been repatriated in 1907, 'the remaining compounds, combined with the Pass Laws and the contract system, were the central institutions of African labour control in South African mining industry' (Van Onselen, 1976, pp. 131-2).

Having started life as a mining camp, Johannesburg in 1887 boasted a total population of 3,000 (Van Onselen, 1986b, p. 2), which by 1904 had grown to number 158,580. Initially, the only employment available on what was later to develop into the Witwatersrand gold complex, was in mining. But already by 1889, with the Witwatersrand total black population numbering between 15,000 and 17,000, only 5,978 blacks were employed in mining. Of the remainder, 1,750 were employed on the construction of the Boksburg tramway (WCM, 1889, pp. 10-11), an infrastructural project necessitated by the east-west extension of the goldmining area. The total permanent population of the Witwatersrand in that year was 30,000, with a 'floating population' of 100,000 (*The Statesman's Year Book*, 1890, p. 923).

Table 3
Witwatersrand and the black mining population, 1889-1904

Year	Number of Black Miners	Total Black Population	Total White Population	Total Wits Population
1889	5,978	16,500	13,500	30,000
1894	40,888	-	-	-
1896	-	51,163 (Jhb)	50,907 (Jhb)	102,070 (Jhb)
1899	96,704	-	-	-
1904 (a)	95,309	74,678 (Jhb)	83,902 (Jhb)	105,580 (Jhb)

Sources: The Statesman's Yearbook; WCM and TCM Annual Reports
(a) - June; (Jhb) - Johannesburg

The gold mines, even though they remained the *raisôn d'être* of the Witwatersrand, was by no means the sole employer for very long. A free market for labour developed around the mines despite the continued existence of the controlled labour-engagement system which characterised Witwatersrand employment when gold-mining was the only form of employment. This system has been undermined right from the word 'go', with the competition for labour between mines, and later merely extended to competition between a multitude of employers covering a whole range of employs.

The extent of this range of employment opportunities for blacks on the Witwatersrand, may be gauged both indirectly, from the number of whites as a proportion of the total population (all employers were white), and directly from the numbers of blacks engaged in other forms of employment, and a comparison of their incomes against that to be had in goldmining. In 1896, Johannesburg, the centre of the Witwatersrand, had a total population of 102,078, of which whites accounted for 50,907 (The Statesman's Year Book, 1901, p. 225). In that same year,

> central Johannesburg was served by: 3,253 domestic servants (mostly black), 3,054 servants (white and coloured), 402 cooks (white), 345 laundresses (coloured), 341 waiters (white), 235 housekeepers (white), 219 nurses, 165 grooms, 146 'houseboys', 106 'kitchenboys', 84 coachmen (black and white), 8 stable-keepers, 5 charwomen, 5 stewards, 4 mother's helps, 3 valets and 1 page (Van Onselen, C., 1982b, p. 3).

The monthly wages for a trained 'houseboy' in 1896 ranged from £3 to £6, in addition to which he received some payment in kind (usually food and clothing). The black mineworker, by contrast, could expect an average income of £3 0s. 10d. (Gool, op. cit., p. 88) But already between 1887 and 1892, an *untrained* 'houseboy' (a 'raw Kaffir') made between £3 and £4 per month (Van Onselen, 1986b, p. 6). This means that this option was immediately open to the newly-arrived mine recruit one year after the fields were declared. By 1904 there were 83,902 whites in Johannesburg out of a total population of 158,580 (The Statesman's Year Book, 1905, p. 246), while by 1906, the town had between 25,000 and 30,000 'houseboys' earning a monthly wage of between £4 and £5 per month, the same as a European housemaid (Van Onselen, 1986b, pp. 9-10).

Particularly instructive is Van Onselen's account of how Africans during the first few months after the war, 'held aloof from domestic service', where cash wages were between £2 10s. and £3 10s. (Van Onselen, 1986b, p. 10). In this context, what chance was there of their freely engaging in mine labour at the then average going rate of £1 13s. per month? (Richardson, op. cit., p. 87) Already by 1899, 'houseboys' were earning an average monthly wage of eighty shillings in contrast to the average fifty shillings per month of black mine workers (Van Onselen, 1986a, p. 18). 'Since the reduction of mine boys' wages there has been a much larger supply of kitchen boys', reported *The Star* after the reduction of black mineworkers' wages in 1896 (quoted in Van Onselen, 1986b, p. 7).

After the end to hostilities in 1902, one may begin to speak of distinctive Witwatersrand black workforce not only as a social entity, but as an entity in political economy. The *de facto* free movement of labour across the Reef meant that the value of black workers' labour-power was beginning to tend towards an average on the Witwatersrand, that average being much higher than the average for the goldmining workforce. At this point one cannot yet speak of manufacturing industry. The tendency of capital to find in its labour-power utilisation the statistically average 'man-hour' (what Ticktin describes as the tendency towards abstract labour) cannot yet be identified. Nevertheless, the bifurcation of the black workforce into a relatively unfree black mine workforce and relatively free non-mine workforce was already a *fait accompli*. There is a great deal of evidence to support a thesis that the eventual construction of two distinct labour forces along racial lines across the entire South African economic landscape, has its origin in this bifurcation of the black workforce at birth.

The goldmining industry, despite the large increase in the Witwatersrand black population, was still wrestling with its old bugbear: labour supply. The mines, having paid labour above its value since the opening of the fields, have consistently tried to reduce black wages to the value of black labour-power. By the end of the South African War, however, the target was beginning to shift. Although the basic problem of getting a large workforce permanently established on the Witwatersrand has been solved, that workforce became increasingly expensive. Geographical distance has been replaced by 'value distance'; demand has been replaced by *effective* demand. The mines could not afford the quantities of black labour it required *because* that labour was free to move, free to seek the highest bidder.

But if the mines could not afford black workers, how could they afford the more expensive Chinese workers? A curious deficiency in Richardson's otherwise very impressive article is his failure to examine Chinese workers' wages. To take the year 1906 (at the height of the Chinese labour experiment): the mines employed 102,420 blacks at £2 12s 3d, and 53,062 Chinese at £2 1s 6d. (Wheatcroft, op. cit., p. 221). The recruiting cost per head per annum for black workers was £4 10s. and for Chinese workers £31 10s., bringing the total cost per head per annum for black workers £142 5s. and for Chinese workers £240 16s.,[37] almost twice as much.

Although the presence of the Chinese workers no doubt contributed to the constant fall in black mine workers' wages from £2 15s. in June 1904, to £2 11s. 11d. in December 1905 (Richardson, op. cit., p. 100), their role cannot be seen merely as one of undercutting. The key lies in their indenture. The indentured labourers, while they cost the mines a great deal more, formed a secure labour force. Their contracts were fixed for three years, and so were their wages and working conditions, and, since they were indentured, so was their employer. In other words, although they were expensive, they had all the attributes which the mines were looking for in its black labour force. The mines' most stubborn problem was the apparently inexorable march towards free labour:

> There was a considerable tendency among African mine workers to dessert from their jobs and escape from the mines. By the time of the Union, in 1910, the annual desertion rate was about 15 per cent.—about 30,000 out of 180,000 (Johnstone, op. cit., p. 37).

This is despite a state of semi-indenture already being *de jure* in place for all black workers in the form of the Masters and Servants

Laws and the Pass Laws. The first made breach of contract by black workers a criminal, rather than a civil offence. The second controlled the movement of black workers, thereby serving as an enabler of the first. Indentured labour allowed the mines to continue production while a mechanism was being set in place for isolating the black mine workforce from the general development of labour at the time. This holds both directly, in that the Chinese compounds became the model for the new closed compounds introduced for black workers after the departure of the Chinese in 1907, and indirectly, in that Masters and Servants Laws and the Pass Laws were made specifically applicable to black mine workers by the Native Labour Regulations of 1911 (ibid., p. 35).

The resort to indentured labour was a retrogressive step inconsistent with the general world development of labour at the time. It was also retrogressive in that that world movement was beginning to express itself in South Africa: a free proletariat was coming into existence directly, paradoxically created by the goldmining industry itself. The goldmining industry, on account of the fixed price of gold, found itself with a foot in both worlds: Fluid international capital seeking a fluid international labour force appeared in goldmining as fluid capital needing to put in place a restricted labour force. Neither on a world scale, nor in South Africa, could restricted labour exist side-by-side with free labour. The undermining of indentured labour as a form came not only from its being, globally speaking, transitional between slavery and wage-labour, but also from its contact with wage-labour in the particular situation. Being in daily working contact with black proletarians, Chinese indentured labourers rioted and deserted as part of their demand, 'for the slackening of their extremely rigid terms of contract, which would allow them at least the same freedom as black labourers' (Kallaway, P. and Pearson, P., 1986, p. 112).[38] Richardson neither explores the relation between free and indentured labour, nor situates these developments on the Witwatersrand within a world context.

Control over the value of labour-power

There can be no question that the significance of the closed compound system in political economy lies in its role as an institution of labour control. Some writers have tried to understand the compounds in terms of the political economy of mining in the entire southern African region (see e.g., Lanning, G., 1979, pp. 173-4 and Van Onselen, 1976, pp. 128-32). But scholarship generally has always understood this control in a limited way. The closed compound system is seen either as a theft-prevention device (especially in diamond mining), or as inhibiting worker organisation, particularly unionisation, or both. These views are not being contested. We wish to argue that these areas of control, important as they were, were, in the case of the goldmining industry, secondary to control over the value of labour-power.

The tendency for free wage-labour to develop amongst blacks on the Witwatersrand despite the non-market way in which mine labour was procured and allocated, presented the mining industry with the problem that its own labour, circulating within a quasi-planning system, was continuously filtering out of that system to fuse into that general body of market-controlled labour. The value of all black labour-power stood to rise dramatically with the acquisition of a host of new needs associated with urban living. Goldmining capital, with the fixed price of its commodity expressing a value much below that of unskilled, urban-based labour-power, could therefore simply not reproduce itself on the basis of such free wage-labour.

The value of the labour-power of its own black workforce had to be restrained. In order to do this, its workforce had to be segregated from the general body of labour. The only competition for labour that goldmining capital would be subject to would be amongst its own members. Such competition at least held out the prospect of being overcome, given the lack of competition to dispose of the product. Under conditions of free wage-labour it would compete with all capitals, including those with commodity-prices consistent with labour-values. The closed compound, damaging as this was to the ability of the Rand to attract labour, was now forced upon goldmining capital by the threat that it might end up with no labour at all. Its reluctance to resort to repression may even be shown by the fact that a daily absentee rate of seldom less than 30% prior to the South African War (Van Onselen, 1976, p. 293n11), did not prompt the mines to demand closed compounds.

Ticktin, too, sees abstract labour as restrained from attaining full maturity in South Africa, and he, too, ascribes this to the peculiar development of labour in the goldmining industry. But Ticktin traces this restraint back to the political events that occurred in the goldmining area in 1922 (Ticktin, op. cit., Ch. 4). Prior to this date, argues Ticktin, South Africa was developing towards normal 'abstract labour'. Subsequently, abstract labour has become 'fractured'. The implication is that prior to 1922,[39] South

Africa had a single, unified workforce, its members free to flow across the economic landscape in search of capital offering the best return, and that this free flow was encouraged by the normal development of capital. We hope to have shown that from the very inception of goldmining, this was never the case. Abstract labour could not become fractured because there never was abstract labour, or, at least, it was *created* already fractured.

The normal process by which the value of labour-power becomes known to capital — that of the lowest wage which will secure the requisite labour-power — was already, in part, foreclosed in the case of black mine labour where the labourers lived in 'open' accommodation on mine property and received food as part of their payment. Every firm has a wage bill and that wage bill is made up of x number of individual wages. Where payments are made in kind for part of what the worker would ordinarily spend out of his wages, those items, although paying towards the value of the worker's labour-power, are no longer part of the wage bill. This is so because they are no longer individually differentiated. Capital procures them as bulk items. Hence African food came under 'stores' in company accounts, and not under 'wages'. It also means that these items could be purchased in bulk — not only for all the workers on one mine, but for the entire industry (see any Chamber of Mines Annual Report). Apart from the obvious cost advantages associated with, on the one hand, bulk buying, and on the other, monopsonistic buying, the most important saving lies in the mines' control over the *standard* of the workers' diet and accommodation, transport, etc. In the 1896 Annual Report of the Chamber of Mines, e.g., we read of the following two resolutions (unanimously adopted) regarding the regulation of compounds:

2. *Re Food* - That the maximum allowance of mealie meal be two and a half pounds per boy per day.

3. That the maximum allowance of meat be two pounds per boy per week.

4. That in addition to the above-mentioned allowance for food, calculated on the number of shifts worked, 25 per cent. be added for boys not working, also that the customary gift of Christmas food be still continued (WCM, 1896, p. 161).

No worker could decide e.g., to travel *to* the mines in seated compartments rather than goods wagons, as these had already been paid for before the worker had even signed on. They did, however, have the option of travelling *back from* the mines by whatever standard of carriage they preferred. As it happened, the standard of carriage which they did prefer was even lower than that transacted by the mines, but that was only because the mines were ignorant of this offer. Once having learnt that the goods-wagon-rate for workers was lower if they were packed in twenty-five to the wagon, the mines wasted no time in adapting to the workers' lower standard and fixing it, and this already in 1897:

> The rate of transport of natives …from Indwe to here [is] 25s. 3d. …The boys on their return to Indwe have the option of engaging a truck; 25 and over can engage a truck, which works out at £1 a head. Now I argue that as a good many boys can go in trucks, and if boys are brought in trucks they should be carried on what I would consider a reasonable truck-load basis. On this basis boys would be brought from Indwe to here for about 16s.

As accommodation and food always make up the largest portion of a worker's expenditure, goldmining capital already had a large measure of control over the worker's capacity to acquire new wants. In 1902, the average cost of housing and feeding a black worker on a Witwatersrand gold mine was £12 5s. 6d. per annum (Charles Goldmann's evidence, ICE, p. 117). But such a worker was still free, nevertheless, to acquire new needs in the remaining areas of his life and to address those needs with money in the wider commercial arena. Although management knew what portion of that value was accounted for by food and accommodation, transport to and from the mines, training and medical attention, there remained an item outside of management's ability to control.

The closed compound brought this unknown factor, too, within management control. Access to prostitutes and alcohol now, also, came under capital's control, although the latter only with a great deal of difficulty.[40] The lack of security for personal property within the open dormitories itself served to restrain any developing need for additional use-values, e.g., shoes, mirrors, etc.[41] and, indeed, compelled the keeping of even the most basic 'historically determined' use-values to a minimum. Thus did the closed compound as a system not only segregate black goldmining labour from the general pool of labour, but also aim to keep the value of labour-power static against the value of labour-power in general. And in this it has succeeded.

Conclusion

In this final section we have been concerned to show that the particular economic history of early South African goldmining and goldmining labour, derives its character not so much from southern African conditions as from world conditions. Theoretical weakness often obscures an *a priori* moralism on the part of scholars, so much so that relevant facts are sometimes ignored, sometimes spuriously interpreted or even denied. This is not a uniquely South African problem, as may be illustrated when Del Mar, who has no theory, makes the correct empirical observation that:

> To-day the mines of Nertschinsk and others in Russia are worked by convicts and serfs; the mines of Mysore are worked by native Indians practically reduced to a condition of slavery the mines of the Witwatersrand, in South Africa, so long as they were worked at all, were worked by 100,000 negros, involuntary labourers, contract-labourers, naked African labourers, bought from their chiefs at so much per head and thrust into the subterranean caverns of Johannesburg to win gold for distant shareholders in London and Paris (Del Mar, op. cit., p. 448).

Although one of the main devices of his book is to consistently draw a distinction between freely-produced gold and gold produced by coerced labour, Del Mar is nevertheless too honest a scholar to 'assist' his project by characterising either the conditions or the form of labour of early South Africa's black mine workers as slavery. In the case of Russia, he is talking of the forms of labour; in the case of India, he is referring to the workers' condition (slavery had been abolished in the British Empire by 1831); while in the case of South Africa he knew that neither the form of labour nor the condition of the labourers could be described as slavery, despite their 'nakedness'. This very 'nakedness' placed them, for Del Mar, outside of the category of American and Australian miners. He solves this problem by, only semi-truthfully, referring to them as, 'involuntary labourers... bought from their chiefs at so much per head'. For the rest, he does what so many scholars have done and continue to do: resort to emotive language and the dramatic phrase. This can never be a substitute for the mortar of theory.

It would appear from this section that the need to maintain a black goldmining labour force separate from the general industrial workforce (regardless of colour) had a far more significant and direct role to play in the eventual creation of an *apartheid* state. Fruitful work might be done in extending this thesis into the present.

Notes

1   For a discussion of the two-fold character of labour, see Marx, 1983, pp. 48-53, but also McMahon, C., 1991, Ch. 3.
2   For a discussion of these questions, see Marx, K., 1977b, Ch. 1.
3   This is taken in the broad sense which includes mental activity.
4   'As soon as the division of labour comes into being, each man has a particular, exclusive sphere of activity, which is forced upon him and from which he cannot escape. He is a hunter, a fisherman, a shepherd, or a critical critic, and must remain so if he does not want to lose his means of livelihood; whereas in communist society, where nobody has one exclusive sphere of activity but each can become accomplished in any branch he wishes, society regulates the general production and thus makes it possible for me to do one thing today and another tomorrow: to hunt in the morning, fish in the afternoon, rear cattle in the evening, criticise after dinner, just as I have a mind, without ever becoming hunter, fisherman, shepherd or critic' (MECW, Vol. 5, p. 47).
5   The starting point of human labour in the abstract is human needs in the abstract.
6   Desai has compiled a collection of Lenin's utterances on economics. Apparently he could not find any thoughts of Lenin on gold (see Desai, M. (ed.), 1989). The fifty volume Lenin Collected Works shows Lenin to have understood nothing more about gold than that it was used in international trade. When orthodox economists mock his extraordinary remark about 'using gold for the building of public lavatories', they do so with some justification.
7   Bonacich, E., 1972, is reviewed under Two Environments and the Question of Free Wage-Labour.
8   Isolated pockets of slavery persisted, but in areas marginal to the world economy. 'Between one-third and one-fifth of the population of West and Central Sudan were slaves at the end of the nineteenth century', according to Sender, J. and Smith, S., 1986, p. 38. See also *Encyclopædia Britannica*, 1982, Vol. XVI, p. 864.

9   For the purposes of this study, there is no need to distinguish between gold occurring in conglomerate beds (Witwatersrand) and ordinary lode gold. For a detailed description of how Witwatersrand gold mining differs technically from that of other regions (see Jeppe, C., 1946a and 1946b).

10 For descriptions of these we have relied especially on Bundy, C., 1979, Hobson, op. cit., Innes, D., 1984, Marks, S. and Rathborne, R. (eds) 1982, Sender, J. and Smith, S., 1986 and Wilson, F., 1972. It is misleading to describe these societies as 'impoverished', as do Crush, J. et al., 1991, p. xiv. The forms in which their wealth could be stored was, however, precarious, rendering them susceptible to the vagaries of nature. 'These societies were subject to cycles of declining productivity and food crises in the face of population change', according to Sender and Smith, op. cit., p. 36.

11 One of the weaknesses of Wolpe's 'modes of production' thesis is that he oversimplifies his characterisation of these economies. Those African economies which did not fit so neatly into his bi-polar scheme of 'African redistributive mode of production vs. capitalist mode of production', such as the short-lived but highly significant peasantries, he simply dismisses as of no importance (see Wolpe, H., 1972, pp. 431-2).

12 For a discussion of the uneven and heterogeneous character of the proletarianisation process in Africa, see Sender, J. and Smith, S, op. cit., esp. Ch. 3. Though the strength of this work lies in its examination of Africa north of South Africa, the general insights are still useful for South Africa.

13 The trickery of and lying to chiefs by private labour touts to induce them to release labour for the Rand was so notorious as to 'spoil the labour market', thereby placing a whole recruiting area out of bounds and compelling mine owners to search further afield. The mines campaigned for the 'putting down' of private touting and called for this function to be performed by properly authorised officials (see Hobson, op. cit., p. 261).

14 'The average stay of a Kafir on the fields is six months, and this is probably too high an estimate...' (WCM, 1890, p. 76). This was particularly irksome to those who would see in it

the racial inferiority of blacks. Sentiments such as 'The Kaffir, with few exceptions, is an unsatisfactory being, who leaves his employment as soon as he gets sufficient money to buy a few wives...', and 'They run away as soon as they have earned enough to get oxen and a wife', were commonplace (see letters to the Chamber reproduced in WCM, 1896, p. 167).

15   It seems rather inappropriate to describe this system as 'debt bondage', as do Richardson and Van Helten, 1982, p. 90. Since so many 'bondsmen' freely absconded and capital 'found themselves compelled' to tolerate this, it was capital, rather than labour, which was in bondage.

16   'This casual to and fro of nearby tribesmen did not satisfy the appetite for labour in the mines', Wheatcroft, op. cit., p. 130.

17   In respect of the pre-mining period, though, this assertion needs to be qualified. These very economies in Mozambique have been sending a regular flow of migrant workers to the Natal colony already from the 1850s (see Innes, op. cit., pp. 50-1).

18   This account is based on Harris, P., 1982. This is an extremely detailed study of the effects of mine labour on the internal dynamics of the African societies in the Delagoa Bay hinterland. Harris, refreshingly, is not afraid to assail some of the sacred cows of South African left wing historiography.

19   This is in sharp contrast to Innes, for whom, 'the process of conquest, ...had created the political and economic conditions for the location of the proletariat in the rural areas rather than around the points of production themselves' (Innes, 1984, p. 60).

20   'The mine compounds, which housed the black workers, were — without exception — situated on mining property', Van Onselen, C., 1982a, p. 5.

21   A point clearly picked up by Hirson, B., 1993.

22   At this particular point Innes was referring to the diamond mines at Kimberley, rather than the gold mines on the Witwatersrand. The substantive point, however, remains.

23   The value of labour-power includes that expenditure which does not take place via the worker's wage, e.g., state-funded education, although, of course, this comes out of surplus-value. In this dissertation we have used the term 'value of

labour-power' without specifying whether or not it includes state expenditure on labour-power reproduction.

24 The political unification and ultimate independence of the four British colonies comprising British South Africa.

25 In the pre-industrial case, exploitation takes many forms, i.a. tribute, slavery, warlordism (or its milder variant, chartered companies), penal colonies (or labour camps), indenture or proletarianisation. In the post-industrial case, the forms include both full and partial proletarianisation, and full and partial de-proletarianisation. None of these, however, by definition implies systematic working to death. Callous murder and systematic extermination did take place, but only extremely rarely was this integral to an economic system. Examples of these are particular kinds of plunder, e.g., those practised by nomadic raiders such as the Mongols (it not being in the interest of settled raiders to destroy the economies they plunder); Stalin's and Hitler's labour camps; and, in all ages prior to the abolition of slavery, mining gold from lode. Only the latter can unequivocally be characterised as 'working to death'; were there is 'no interest in preserving the lives of the workers who were then often [in this case, always] paid below value', to borrow Ticktin's phrase.

26 Trewhela falls foul of his own prescription: 'Least of all, therefore, can the money-commodity segregate itself from the market behind ghetto walls, any more than a king remains king without subjects, or the Good Shepherd tend sheep without Christians' (ibid., p. 18).

27 'The conflict... takes place between the marketable value demanded by the supplier and the marketable value supplied by the demander' (MECW, Vol. VI, p. 118).

28 This is not a uniquely southern African phenomenon, as Bonacich explains:

> The characteristic feature of the labor market in most of Africa has always been the massive circulation of Africans between their villages and paid employment outside. In some places villagers engage in wage-earning seasonally. ...The African villager, the potential migrant into paid employment, has a relatively low, clearly defined and rigid income goal; he wants money to pay

head and hut taxes, to make marriage payments required of prospective bridegrooms, or to purchase some specific consumer durable (a bicycle, a rifle, a sewing machine, a given quantity of clothing or textiles, etc.) (Bonacich, E., op. cit.).

29 ICE, pp. 13-14. 'If wages [of workers who work only to amass a certain amount of money] were to rise, workers would reach their desired income and withdraw more quickly from the market' (ibid.).
30 Although Hobson makes the same distinction, he argues it slightly differently (see pp. 257-8, 266, where this device is discussed as integral to colonial land dispossession and proletarianisation, and pp. 268-72 for its examination in South Africa).
31 According to Hobson a *corvée* was imposed in Natal (ibid., p. 256).
32 One is reminded of the physicist's conundrum: the very act of determining the position of an electron, alters its position.
33 Van Onselen's well-researched and richly-anecdoted two-volume study of the social and economic history of the early Witwatersrand adds a welcome human dimension to these questions.
34 'In British Guiana the local labor force, composed mainly of African ex-slaves, called a series of strikes in 1842 and 1847 against planters' attempts to reduce their wages. Plantation owners responded by using public funds to import over 50,000 cheaper East Indian indentured workers. A similar situation obtained in Mississippi, where Chinese were brought in to undercut freed blacks' (Bonacich, op. cit., p. 554).
35 According to Van Onselen, the Cinderella Prison, opened on the East Rand in 1905, 'supplied local mines with convict labour' (Van Onselen, C., 1982b, p. 179).
36 Richardson, P., 1977, p. 86. The idea, however, has already been mooted at least since 1896 (see letter of Markham to Chamber of Mines, 19 February 1896, reproduced in WCM, 1896, p. 167).
37 Calculated from figures provided in Richardson, op. cit., p. 93n28. Black workers remained on the mines for periods ranging from six to fifteen months. Our calculations are based on one year and we have assumed that all workers were

fresh recruits, ignoring returnees, who would be slightly cheaper.

38 A poster of the time, depicting an immaculately attired African, complete with cigar and bicycle, in conversation with two haggardly dressed Chinese, explains Chinese desertion with the caption, 'Allee same us wantee be like Kaffir' (ibid., p. 118).

39 The defeat of the Rand Revolt of 1922 also signifies, for Ticktin, the beginning of racial discrimination as a system in South Africa. The controversy surrounding this position lies outside of the scope of this thesis.

40 (See the controversy over the Liqour Law in ICE, *passim*, and the chapter 'Randlords and Rotgut' in Van Onselen, 1982a).

41 While outside the compounds, 'People become locked into the world economy as former luxury imports become necessities and that this was an important factor in generating migrant labour. It was not merely human nature, however, that encouraged the importation of goods like cloth and liquor...' (Harris, op. cit., p. 157).

Conclusion

This study has attempted to demonstrate that the manner in which gold plays its economic role was transformed in the second half of the nineteenth century. A theory of gold derived from its history prior to this transformation, such as Marx's, could, like Aristotle's theory of value, necessarily only attain to partial completion. One of the objects of this thesis has been to contribute towards the completion of such a theory.

There are two deficiencies in Marx's theory of gold. The first relates to the contradictory demands of commodity and money, and of money and capital on this material. The second relates to the value-relationships within the actual production process of gold, and the changes in those value-relationships. This dissertation has attempted to address these deficiencies.

We have aimed to achieve this by re-examining the nineteenth century in detail, and described the relevant developments which have taken place. These fall into two broad categories: those relating to the circulation of gold (gold as money); and those relating to its production (gold as use-value). In this dissertation, Marx's treatment of the circuit of capital in gold production, as presented in Volume II of *Capital*, was examined in the light of these developments. It is our contention that Marx's discussion of gold production is undeveloped. Similarly, he does not have an adequate picture of the nature of gold-producing labour-power. In Volume I of *Capital*, as well as in *A Contribution to the Critique of Political Economy* and in the *Grundrisse*, gold is examined *as money*, hence his characterisation of gold-producing labour as necessarily oppressive where gold embodies independent exchange value. The circuit of gold-producing capital in Volume II, though, assumes gold *as capital*. But gold is not empirically examined as capital, only as money. The character of gold-producing labour is assumed to remain unchanged when, in fact, it requires re-examination. The theory is in need of refinements which Marx, being either unaware of these developments or not in a position to assimilate them before the end of his life, was unable to carry out himself. Hence Engels' lengthy footnote in Volume II of *Capital*.

How Marx's thoughts on these questions might have developed remains unclear. His recently published notebook *Bullion: das vollendete Geldsystem*, is itself in a very *unvollendete* state (MEGA, Vol. 8, IV/8, pp. 767-73). The manuscript *Geldwesen, Kreditwesen, Krisen* remains unpublished. The original, housed in the Institute of Social History (now the Marx-Engels Foundation) in Amsterdam, we have been unable to study for the illegibility of Marx's handwriting (Manuscript B-72 of the Marx-Engels Nachlaß collection, Institute of Social History, Amsterdam).

But one lesson which emerges very strongly from this study is that the scholar has to keep a clear distinction between gold as use-value, gold as money and gold as capital, while, at the same time, always seeing these distinctions in their mutual interrelation. This, e.g., is where Trewhela might have distinguished himself from Williams, had he not decided to leave out of account gold as tooth-filling, etc.

In this dissertation we have devoted a substantial amount of attention to those scholars who have tried to show a direct relationship between the social role of gold, and the character of the labour which produces it. Although we have taken issue with both Williams and Trewhela, they distinguish themselves from Innes in that they have been creative, whereas Innes, without demonstrating any knowledge of Marx's thought processes, offers Marx's conclusions as axioms. Both Williams and Trewhela are correct in the sense that they see the social role of gold and the labour which produces it as intimately related. This is not appreciated in the literature generally. The narrative of this thesis consisted in tracing the course of this intimacy.

It is the social character of gold in relation to the social character of the labour which produces it, which is mediated in the particular situation by the concrete character of gold in relation to the concrete character of the labour which produces it. This is how we have tried to understand the complex relation between South African gold and South African gold-producing labour. In the light of this, early South African goldmining history does not appear to be the aberration it is so often made out to be. Its character was consistent with a world pattern at the time. Late nineteenth century goldmining, in the South African context, took the form of South African goldmining.

This is by no means to suggest that South African conditions were free from evil and wrongdoing. Racism and abuse there were. Brutal mine foremen and unscrupulous recruiters, too. Racist and corrupt border officials so often robbed Africans of their earnings that the mines were eventually moved to intervene, as they were when the white public of Johannesburg became alarmed at the number of dead bodies of black mine workers left on the streets to rot. A distillery, *De Eerste Fabrieken in de Zuid Afrikaansche Republiek* (such, proudly, was its actual name), was the instrument for turning large numbers of mine workers into alcoholics on potato whiskey. Sufficient evidence exists for a most horrific picture of abuse to be sketched.

Up to the advent of generalised wage-labour, the history of goldmining is a history written in blood. No particular individual or class is responsible for humanity having paid so grotesque a price. It is the price paid by humanity for its own social development. Goldmining only becomes working to death once gold becomes money. That particular transformation of its social character is only a signature of the transformation of the social character of humanity. The scholar's humanity is not diminished by a refrain from expressing outrage at goldmining as such. Neither does goldmining become more humane by the scholar inveighing against its horrific human cost. Human labour itself, of which gold is the material hand maiden, has to develop beyond the need for a value form.

We have deliberately avoided questions of ethics and morality. The reason for this is that we have seen the scientific credentials of too many works compromised by the writer's moral outrage at some or other perceived abuse. The most obvious is the difference in wages between black and white mine workers on the Witwatersrand. To most writers this was discrimination — they should be paid the same. Closely related to this is the question of the actual level at which African mine workers were paid. Sentences in which actual figures are given, are sometimes punctuated with exclamation marks. Very few scholars have approached wages for what they are: the money equivalent of the labourer's *necessary* consumption. Sometimes this point is very tentatively touched on and then immediately backed away from. Gool even goes so far as to discuss differential value of labour-power in detail, but then dismisses his own dispassionate examination with, 'This provides a marvellous rationale for low wages' (Gool, op. cit., p. 27). The same can be said of some of the international works. Del Mar, e.g., inveighs at length against the role of slavery and other forms of unfree labour in the history of gold-production, despite himself producing enough material to actually explain it. As for the slave trade, we were not interested in cataloguing its horrors, but in understanding it as the starting point for universalised wage-labour. To discuss wages, even extremely low wages, in a scientific manner does not imply approval of discrimination or under-paying. It indicates only an attempt to see an object for what it really is, rather than to have that attempt compromised by how one might feel about it.

Another area where morality, or *political* morality, interferes with scientific examination is in respect of the role and nature of capital. In leftwing South African historiography, there is a tendency for capital to be automatically dismissed as inherently incapable of anything positive. Thus, for example, is the very important 1897 *Report of the Industrial Commission of Enquiry into the Goldmining Industry* of the then South African Republic rejected out of hand by many writers as 'industry propaganda', especially as the ZAR government did not have the expertise to oppose the claims of the engineers and managers of the Chamber of Mines.

Similarly, capitalists are only quoted when they incriminate themselves, i.e., confirm their nastiness. Their scientific interests (subject to the needs of capital, of course) are simply not noticed. It is often the case that capital has a more acute perception of a question than have others, simply because it stands to loose money if it did not. It cannot be denied that the 1897 report was prepared by mining capital for mining capital, but does that necessarily render it devoid of all truth, not worthy of academic scrutiny? We have found only one writer who has made a serious study of this critical document. Like this isolated scholar, we have not shied away from turning to capital when it was the source of useful material.

It has not been our object either to, 'place the African working-class more in the centre of the stage', or to expose the collusion between capital, the state and white workers against black workers. One scholar regards it as, 'essential that we place the class struggle to the fore', without establishing just how backward or forward that struggle actually was (Williams, op. cit., p. 7). The object has been to study gold and the labour which produces it. That is a universal question of which South Africa is a particular expression. We hope that we have managed to demonstrate, e.g., that the eventual closed compound system of the gold mines, which is so often automatically ascribed to management's need to contain unionisation (of course, it did have that effect), relates to a *universal* problem of the need to restrain the value of gold-producing labour-power below that represented by the price of gold. We hope that we have shown, further, that without a study of the particular form assumed by competition amongst gold-producing capitals, it becomes very difficult to explain why South African goldmining capital stopped short of outright monopoly and put in place, instead, the group system. This being despite the monopoly in diamond mining, and the 'monopoly stage' of capitalism.

This dissertation has considered the relation between gold-producing capital and South Africa's black goldmining workforce, in preference the white workforce or the total workforce, for a number of reasons: (i) the nature of gold as commodity-capital, as distinct from other commodities as commodity-capital, and the nature of gold-producing capital as distinct from capital in general, emerges far more strongly where gold-producing capital itself has the opportunity to structure the capital-labour relationship, rather than inheriting it from history, as with the white workers; (ii) white workers and capital related to each other in the 'classical' manner, which means that this relationship has already been explained by, i.a., Marx; (iii) scholarship has failed to appreciate that the relation between gold-producing capital and black labour describes another *form* of capital-labour relation, rather than merely *a greater degree of antagonism* within the same form. This thesis aimed at providing a different basis for understanding these relations.

To conceptualise the determination of early goldmining capital to reduce the pay of black goldminers in South Africa as a wages struggle is erroneous. The concern of capital was far more fundamental. In its own parlance, capital was concerned with the 'cost of living' of its workforce. Ordinarily this cost of living, a definite portion of the value of labour-power, is a real, calculable figure. Its computability involves a simple addition of all the worker's (and his family's) expenses. This presupposes that the worker's necessaries of life take the form of *commodities*. By itself, however, this precondition is only adequate for revealing to the worker the value of his labour-power. Capital, too, could formally know this simply by asking the worker what his weekly or monthly expenses are. But this is inconsistent with the capital-labour relation, as Williams illustrates by his strange complaint that capital did not give the workers 'a modicum of say' in determining the value of their labour-power (when, in fact, he means the level of their wages). Capital does not pay for labour, but for labour-power — it hires the worker's capacity to labour for a definite period of time. But it has, in fact, no real interest in the value of labour-power. The value of labour-power forces its way into capital's calculations as a necessary deduction from the total value. In this way does capital know the value of labour-power as the complement of its surplus-value. Wage struggles are struggles over this *known figure*.

In early South Africa goldmining, only white workers found themselves in such a relation with capital. The value of white workers' labour-power was known to both workers and capital down to the minutest detail. This is the basis of capital's inability to reduce white workers' wages, rather than the fact that they were skilled or had trade unions (although, of course, these did play an important role). When capital did finally assault white workers' wages, it was assaulting their *livelihood*. To talk of workers' resistance and unionisation is appropriate in this respect.

The same cannot be said for black workers. Not only was the bulk of the workers' necessaries of life not commodities, and hence not expressible in value terms, but as such they circulated outside of the circuit of capital. That is to say, capital in its relation to black workers had no unified circuit. It was painfully obvious to capital that the workers were retaining part of the *surplus*-value for themselves, something which 'rightfully' belongs to capital.

Any talk of 'organisation', 'resistance', 'bargaining power', 'union rights', etc., with reference to black goldminers in this early period is, therefore, inappropriate. They were not engaged in a capital-labour relation since capital could not properly function as capital and labour could not become fully wage-labour. Having bought the worker's labour-power, capital was but its formal owner, not its effective owner, not its possessor. Similarly could labour-power not properly be a commodity, since capital was unable to express its exchange value for it.

There are two ways of solving this kind of problem (from the point of view of capital). This could be done either by forcing the workers 'into the cash nexus', or by simply offering less in exchange for their labour-power. Why this was not so straightforward a matter and how it was pursued have been discussed in detail.

In the case of goldmining capital in particular, finding the definite figure which describes the value of labour-power (or, from the point of view of capital, the necessary deduction from value) assumes a particular urgency on account of the fixed price of the product. Regardless of how large or small the magnitude of value objectified in that product, it will only ever be transformed into a predetermined value-total of other commodities.

In theory this does mean that if gold could be produced at a *lower* value than that represented by its fixed price, the goldmining industry would be drawing a rent from the rest of society. But producing gold at a *higher* value does not mean that, in turn, society draws a rent from goldmining. Instead, the value of labour-power in goldmining is reduced and kept low, so that in effect, society draw the rent from the goldmining workforce directly. Goldmining *capital* escapes this rent.

In the particular situation of South Africa, it is the existence of two workforces whose respective reproduction related to the circuit of capital in two very different ways — that circuit running under the particular value-constraint of the fixed gold price — that gives the political economy of *South African* gold its historically peculiar character.

Restriction of the value of black goldminers' labour-power by South African gold-producing capital on the one hand communalises the restrictions free gold-diggers individually placed on the value of their own labour-power, and, on the other, transforms the working to death of pre-capitalist goldmining into a value-relation.

What remains to be done is to account for the twentieth century history of gold in terms of the same set of concerns as was done for the period reviewed in this thesis. One and the same thing, the material gold, is at the same time two different commodities, i.e., two different contradictions. The history of gold in the twentieth century, the evidence seems to suggest, is the working out of the contradiction between these two contradictions, i.e., between commodity I (the first contradiction) and commodity II (the second contradiction). So commonplace has gold as commodity II become, that when one today speaks of gold as a commodity, it is automatically understood as an investment opportunity, rather than as a 'thing of utility' which has an exchange value. Gold as capital has entirely eclipsed gold as money, which, in turn, has eclipsed gold as use-value. The contradiction between capital and money, and the role of gold in world credit economy, need to be examined. This must be the next chapter in the development of a theory of gold.

About the social character of gold we may conclude that the same physical substance, gold, occurs simultaneously in a variety of social forms. In its physical consistency it is the one product in nature most adequate to the physical consistency of labour — as productive human activity. Gold, in whatever social form, remains gold; so does labour, whatever social form it may take, remain labour. Gold, in whatever concrete form it is fashioned, remains gold; labour, in whatever concrete form exercised, remains labour. Gold is the compass by which labour negotiates its way through its troublesome adolescence as value. The history of gold is, in the most literal sense then, the history of the making of humankind. Only upon his labour maturing as freely-associated labour can that labour eventually lay its golden lodestone down.

Bibliography

Ally, R. (1990), *The Bank of England and South Africa's Gold, c.1886-c.1926*, PhD. thesis, University of Cambridge.

Anikin, A. (1983), *The Yellow Devil: Gold and Capitalism*, Progress Publishers: Moscow.

Asiegbu, J. (1969), *Slavery and the Politics of Liberation 1787-1861. A Study of Liberated African Emigration and British Anti-slavery Policy*, Longmans: London.

Bardo, M. and Schwartz, A. (eds.) (1984), *A Retrospective on the Classical Gold Standard, 1821-1931*, Univ. of Chicago Press: Chicago.

Beilharz, E. A. and López, C. U. (eds.) (1976), *We Were 49ers! Chilean accounts of the California Gold Rush*, Ward Ritchie Press: Pasadena, California.

Bernstein, E. (1958), 'The Gold Crisis and the Gold Standard' in *Quarterly Review and Investment Survey*, 1958, pp. 1-12.

Berton, P. (1958), *Klondike: The Life and Death of the Last Great Goldrush*, McClelland and Stewart, Toronto.

Bhappu, R. and Lewis, F. (1975), 'Gold Extraction from Low Grade Ores: Economic Evaluation of Processes' in *Mining Congress Journal*, January 1975, pp. 38-41.

Blainey, G. (1965), 'Lost Causes of the Jameson Raid' in *Economic History Review*, 2nd ser., Vol. XVIII.

_____ (1978), *The Rush that Never Ended: A History of Australian Mining*, Melbourne.

Bloomfield, A. (1963), *Monetary Policy under the International Gold Standard, 1880-1914*, FRBNY: New York.

_____(1963), 'Short-Term Capital Movements under the Pre-1914 Gold Standard', *Princeton Studies in International Finance*, no. 31, Princeton Univ. Press: Princeton.

Bonacich, E. (1972), 'A Theory of Ethnic Antagonism: The Split Labor Market' in *American Sociological Review*, Vol. XXXVII, no. 5, (Oct), pp. 547-59.

Brewer, W. M. (1895), 'The Gold-Regions of Georgia and Alabama' in *Transactions of the American Institute of Mining Engineers*, Vol. XXV, Feb. 1895-Oct. 1895, pp. 569-587.

Brown, B. (1978), *Money Hard and Soft*, Macmillan: London.

Brown, M. (1989), 'Third Great Wave of Mine Development' in *South African Mining: Coal, Gold and Base Metals*, May 1989, pp. 17-25.

Brown, W. (1940), *The International Gold Standard Reinterpreted, 1914-1934*, 2 Volumes, National Bureau for Economic Research: New York.

Bundy, C. (1979), *The Rise and Fall of the South African Peasantry*, Heinemann: London.

Buranelli, V. (1981), *Gold: An Illustrated History*, Book Club Associates: London.

Burckhardt, W. (1886), *The Currency Problem: A Proposal for the Rehabilitation of Silver*, Effingham Wilson: London.

Busschau, W. (1949), *Measure of Gold*, Central News Agency: Johannesburg.

Cardwell, D. (1972), *Turning Points in Western Technology*, Neale Watson: New York.

Cartwright, A. (1967), *Gold Paved the Way: The Story of the Goldfields Group of Companies*, Macmillan: London.

Cendrars, B. (1982), *Gold*, Peter Owen: London.

Chandler, A. (1990), *Scale and Scope: The Dynamics of Industrial Capitalism*, Belknap Press: Cambridge, Mass.

Chevalier, M. (1859), *On the Probable Fall in the Value of Gold, the Commercial and Social Consequences which May Ensue and the Measures which it Invokes*, Alexander Ireland: Manchester.

Clapham, J. (1961), *The Economic Development of France and Germany, 1815-1914*, Cambridge University Press.

Clark, W. (1970), *Gold Districts of California*, (Bulletin 193), California Division of Mines and Geology: San Francisco.

Cochrane, P. (1980), 'Gold, The Durability of a Barbarous Relic', in *Science and Society*, no. 4.

Conquest, R. (1978), *Kolyma: The Arctic Death Camps*, Macmillan: London.

Cooper, R. (1982), 'The Gold Standard: Historical Facts and Future Prospects', *Brookings Papers on Economic Activity*, no. 1, pp. 1-45.

Crowther, G. (1943), *An Outline of Money*, Thomas Nelson & Sons: London.
Crush, J. (1987), 'Restructuring Migrant Labour on the Gold Mines' in *South African Review*, no. 4, Ravan Press: Johannesburg, pp. 283-91.
Crush, J, et al. (1991), *South Africa's Labour Empire: A History of Black Migrancy to the Gold Mines*, David Philip: Cape Town.
Curle, J. (1902), *Gold Mines of the World*, Waterlow and Son.
Davis, D. (1984), *Slavery and Human Progress*, OUP.
Day, A. (1957), *Outline of Monetary Economics*, OUP: London.
Day, J. (1978), 'The Great Bullion Famine of the Fifteenth Century' in *Past and Present*, no. 79, pp. 3-54.
De Brunhoff, S. (1973), *Marx on Money*, Urizen Books: New York.
De Cecco, M. (1984), *The International Gold Standard: Money and Empire*, Francis Pinter: London.
De Quille, D. (1969), *The Big Bonanza: An Authentic Account of the Discovery, History and Working of the World-Renouned Comstock Lode of Nevada*, Eyre & Spottiswoode: London.
Del Mar, A. (1969), *A History of the Precious Metals*, Augustus M. Kelley: New York.
Denoon, D. J. N. (1967), 'The Transvaal Labour Crisis, 1901-6' in *Journal of African History*, Vol. VII, no. 3, pp. 481-494.
Desai, M. (ed.) (1989), *Lenin's Economic Writings*, Lawrence and Wishart: London.
Douglas, H. (1987), *The Impact of Technological and Social Change on Mineral Exploration, ... etc.*, MPhil Thesis, Imperial College of Science and Technology (Dept of Geology). British Libr. no. TXXL 81349.
Draper, A. (1979), *Operation Fish: The Race to Save Europe's Wealth, 1939-1945*, Cassell: London.
Duffy, I. (1982), 'The Discount Policy of the Bank of England during the Suspension of Cash Payments, 1797-1821' in *Economic History Review*, 2nd ser., Vol. XXXV, pp. 67-82.
Dunning, J. (1988), *Explaining International Production*, Unwin Hyman: London.
Du Plessis, A. (1989), 'Water Hydraulically Powered Stoping: An Opportunity for Significantly Improving Mining Profits' in *South African Mining: Coal, Gold and Base Metals*, May 1989, pp. 42-51.
Eaton, L. (1948), 'Metal Mining' in *75 Years of Progress in the Mineral Industry*, American Institute of Mining Engineers: New York.

Eichengreen, B. (ed.) (1985), *The Gold Standard in Theory and History*, Methuen: London.

Eltis, D. and Walvin, J. (1981), *The Abolition of the Atlantic Slave Trade. Origins and Effects in Europe, Africa, and the Americans*, University of Wisconsin Press.

Emden, P. (1938), *Money Powers of Europe in the Nineteenth and Twentieth Centuries*, Appleton-Century: New York.

Encyclopædia Brittanica (1982), Vol. VIII (pp237-240), Vol. XII (pp245-257), Vol. XVI (pp853-866).

Erskine, J. (1852), *A Short Account of the Late Discoveries of Gold in Australia, With Notes of a Visit to the Gold District*, T & W Boone: London.

Evans, D. (1969), *The History of the Commercial Crisis, 1857-58 and the Stock Exchange Panic of 1859*, (reprint), Augustus M. Kelley: New York.

Finley, J. (1910), *The Cost of Mining: An Exhibit of the Results of Important Mines throughout the World*, McGraw-Hill: London.

Fordyce, W. (1924), *In Search of Gold*, T. C. & E. C. Jack: London.

Frankel, S.: (1936), 'Fifty Years on the Rand', in *Economist*, 19 September.

―――― (1967), *Investment and the Return to Equity Capital in the South African Gold Mining Industry, 1887-1965—An International Comparison*, Harvard Univ. Press: Cambridge, Mass.

Fraser, M. and Jeeves, A. (1977), *All that Glittered: Selected Correspondence of Lionel Phillips, 1890-1924*, Oxford Univ. Press: Cape Town.

Frieden, J. (1989), *Banking on the World: The Politics of International Finance*, Hutchison Radius: London.

Friedman, M. and Schwartz, A. (1963), *A Monetary History of the United States, 1867-1960*, NBER, New York, Princeton Univ. Press: Princeton, N. J.

Gool, S. (1983), *Mining Capital and Black Labour in the Early Industrial Period in South Africa: A Critique of the New Historiography*; Skrifter Utgivna av Ekonomisk-Historiska Föreningen: Lund.

Graham, F. and Whittlesey, C. (1978), *Golden Avelanche*, (reprint), Arno Press: New York.

Gray, J. (1937) *Payable Gold: An Intimate Record of the History of the Discovery of the Payable Witwatersrand Gold Fields and of Johannesburg in 1886 and 1887*, Johannesburg.

Green, T. (1968), *The World of Gold*, Michael Joseph: London.

_____(1971), 'Gold Smuggling in the Seventies', in *Euromoney*, August 1971.
_____(1985), *The New World of Gold*, Weidenfeld and Nicholson: London.
_____(1987), *The Prospect for Gold: The View to the Year 2000*, Walker: New York.
Greever, W. S. (1963), *The Bonanza West. The Story of the Western Mining Rushes, 1848-1900*, University of Oklahoma Press.
Gregory, C. (1980), *A Concise History of Mining*, Pergamon Press.
Gregory, T. (1932), *The Gold Standard and its Future*, E. P. Dutton: New York.
Hacche, J. (1970), *Economics of Money and Income*, Heinemann: London.
Harris, P. (1982), 'Kinship, Ideology and the Nature of Pre-Colonial Labour Migration' in Marks, S. and Rathbone, R., *Industrialisation and Social Change in South Africa*, Longman: New York, pp. 142-66.
Hatch, F. and Chalmers, J. (1895), *The Gold Mines of the Rand, being a Description of the Mining Industry of the Witwatersrand Republic*, Macmillan: London.
Helfferich, K. (ed.) (1900), *Ausgwählte Reden und Aufsätze über Gold- und Bankwesen*, (Schriften des Vereins zum Schutz der deutsche Goldwährung, Bd. I), Guttenberg: Berlin.
Hellemans, A. and Bunch, B. (1988), *The Timetables of Science: A Chronology of the Most Important People and Events in the History of Science*, Simon & Schuster: New York.
Helper, H. (1948), *Dreadful California*, The Bobbs-Merrill Company, Indianapolis: New York.
Hinshaw, B. (ed.) (1967), *Monetary Reform and the Price of Gold*, Johns Hopkins Univ. Press: Baltimore.
Hirson, B. (1993), 'The Trotzkyists of South Africa, 1932-48' in *Searchlight South Africa*, no. 10, April 1993.
Hobart Houghton, D. (1973), *The South African Economy*, OUP: London.
Hobson, J. A. (1938), *Imperialism*, George Allen & Unwin Ltd: London.
Hoffman, A. (1946), *Free Gold: The Story of Canadian Mining*, Rinehart & Co.: Toronto.
Hyndman, H. (1967), *Commercial Crisis of the Nineteenth Century*, Augustus M. Kelley: New York.
Innes, D. (1981), 'Capitalism and Gold' in *Capital and Class*, Vol XIV, pp. 5-35.

―――― (1984), *Anglo: Anglo American and the Rise of Modern South Africa*, Ravan: Johannesburg.

Investors' Guardian (1910), *Reduction of Working Costs on the Rand*, Investors' Gaurdian, London.

Jacob, W. (1968a), *An Historical Inquiry into the Production and Consumption of the Precious Metals*, vol. I, Augustus M. Kelley: New York.

―――― (1968b), *An Historical Inquiry into the Production and Consumption of the Precious Metals*, vol. II, Augustus M. Kelley: New York.

Jamieson, B. (1990), *Goldstrike!: The Oppenheimer Empire in Crisis*, Hutchison: London.

Jastram, R. (1977), *The Golden Constant: The English and American Experience, 1560-1976*, John Wiley & Sons: Chichester.

―――― (1981), *Silver: The Restless Metal*, J. Wiley: New York.

Jeeves, A. (1985), *Migrant Labour in South Africa's Mining Economy: The Struggle for the Gold Mines' Labour Supply, 1890-1920*, McGill-Queens University Press: Kingston and Montreal.

Jeppe, C. B. (1946a), *Goldmining on the Witwatersrand*, Volume I, Tvl Chamber of Mines, Cape Times Ltd: Cape Town.

―――― (1946b), *Gold Mining on the Witwatersrand*, Volume II, Tvl Chamber of Mines, Cape Times Ltd: Cape Town.

Johnson, B. (1970), *The Politics of Money*, John Murray: London.

Johnson, P. (1987), *Consolidated Gold Fields: A Centenary Portrait*, Weidenfeld and Nicolson: London.

Johnstone, F. (1976), *Class, Race and Gold: A Study of Class Relations and Racial Discrimination in South Africa*, RKP: London.

Joughin, N. (1978), 'Progress in the Development of Mechanised Stoping Methods' in *Journal of the South African Institute of Mining and Metallurgy*, March 1978, pp. 207-217.

Kallaway, P. and P. Pearson (1986), *Johannesburg: Images and Continuities. A History of Working Class Life Through Pictures 1885-1935*, Raven Press: Johannesburg.

Katzen, L. (1964), *Gold and the South African Economy*, Balkemu: Cape Town.

Kemmerer, D. (1975), 'The Role of Gold in the Past Century' in Sennholz, H. (ed.) (1975), *Gold is Money*, Greenwood Press: London, pp. 104-121.

Kemp, T. (1969), *Industrialisation in Nineteenth Century Europe*, Harlow.

Kent, W. (ed.) (1985), *Money Talks: The 2500 Greatest Business Quotes from Aristotle to DeLorean*, Facts on File Publications: Oxford, UK.

Kessler, L. (1904), *The Gold Mines of the Witwatersrand and the Determination of their Value*, Edward Stanford: London.

Kettel, B. (1982), *Gold*, Graham & Trotman: London.

Kindleberger, C. (1987), *A Financial History of Western Europe*, George Allen and Unwin: London.

King, J. (1977), *A Mine to Make a Mine: Financing the Colorado Mining Industry, 1859-1902*, Texas A & M University Press.

Kritz, M. (1967), 'Gold: Barbarous Relic or Useful Instrument?' (Essays in International Finance, No. 60, June 1967), International Finance Section, Department of Economics, Princeton University: Princeton, New Jersey.

Kubicek, R. (1972), 'The Randlords in 1895: A Reassessment' in *Journal of British Studies*, Vol. XI, pp. 84-103.

_____(1979), *Economic Imperialism in Theory and Practice: The Case of the South African Gold Mining Finance, 1886-1914*, Duke Univ. Press: Durham.

Landes, D. (1981), *The Unbound Prometheus: Technological Change and Industrial Development in Western Europe from 1750 to the Present*, Cambridge University Press.

Lanning, G. and Mueller, M. (1979), *Africa Undermined: Mining Companies and the Underdevelopment of Africa*, Harmondsworth: Penguin.

Legassick, M. (1983), 'Gold, Agriculture and Secondary Industry in South Africa, 1885-1970: From Periphery to Sub-Metropole as a Forced Labour System' in Palmer, R. and Parsons, N. (eds), *The Roots of Rural Poverty in Central and Southern Africa*, Heinemann: London, pp. 175-200.

Leith, C. (1931), *World Minerals and World Politics*, McGraw Hill: New York.

Lenin, V.I. (1921), *The Importance of Gold now and after the Complete Victory of Socialism*, in LCW, Vol. XXXIII, pp. 113-4.

_____ (1978), *Imperialism, Highest Stage of Capitalism*, Progress: Moscow.

Letcher, O. (1936), *The Gold Mines of Southern Africa: The History, Technology and Statistics of the Gold Industry*, Waterlow & Sons: London.

Lindley, C. (1897), *A Treatise on the American Law Relating to Mines and Minerals Lands within the Public Land States and Territories and Governing the Acquisition and Enjoyment of Mining Rights in Lands of the Public Domain* (Vol.1/2), Bancroft-Whitney Co.: San Francisco.

Lipton, M. (1986), *Capitalism and Apartheid: South Africa, 1910-1986*, Wildwood House, Aldershot: England.

Lock, A. (1882), *Gold: Its Occurrence and Extraction*, E. & F. N. Spon: London.

Mandel, E. (1971), *Marxist Economic Theory*, The Merlin Press: London.

_____(1974), *The Decline of the Dollar: A Marxist View of the Monetary Crisis*, Monad: New York.

Marks, S. and Rathbone, R. (1982), *Industrialisation and Social Change in South Africa*, Longman: New York.

Marks, S. and Trapido, S. (1979), 'Lord Milner and the South African State' in *History Workshop*, No. 8, Autumn, pp. 50-80.

Marx, K. (1969), *Theories of Surplus Value*, Part 2, Lawrence and Wishart: London.

_____(1972), *Theories of Surplus Value*, Part 3, Lawrence and Wishart: London.

_____(1973), *Grundrisse*, Penguin: Harmondsworth.

_____(1976), *Economic and Philosophic Manuscripts of 1848*, International Publishers: New York.

_____(1977a) *Capital*, Vol. II, Lawrence & Wishart: London.

_____(1977b) *Contribution to the Critique of Political Economy*, Progress: Moscow.

_____(1983), *Capital*, Vol. I, Lawrence & Wishart: London.

_____(1984), *Capital*, Vol. III, Lawrence & Wishart: London.

_____(1985a) *Wage Labour and Capital*, Progress Publishers: Moscow.

_____(1985b), *Wages, Price and Profit*, Progress Publishers: Moscow.

Mathieson, W. (1967), *British Slave Emancipation 1838-1849*, Octagon Books Inc.: New York.

McGrane, R. (1965), *The Panic of 1837: Some Financial Problems of the Jacksonian Era*, University of Chicago Press: Chicago and London.

McMahon, C. (1991), *Marx and Hegel: Methodology and the 'Historical Transformation Problem'*, PhD. thesis, University of Glasgow.

Meikle, S. (1985), *Essentialism in the Thought of Karl Marx*, Duckworth: London.

Moggridge, D. (1969), *The Return to Gold, 1925: The Formulation of the Policy and its Critics*, Cambridge Univ. Press: Cambridge, UK.

Molesworth, G. (1891), *Silver and Gold: Money of the World*, Bimetallic League of England.

Morgan, E. (1976), *Gold or Paper?: An Essay on Governments' Attempts to Manage the Post-war Monetary System and the Case for and against Restoring the Link to Gold*, Institute of Economic Affairs, London (Hobart Papers no. 69).

Nitze, H. and Wilkens, A. (1895), 'The Present Condition of Gold Mining in the Southern Appalachian States' in *Transactions of the American Institute of Mining Engineers*, Vol. XXV, Feb. 1895-Oct. 1895, pp. 661-795.

Pallister, D., Stewart, S. and Lepper, I. (1988), *South Africa Inc.: The Oppenheimer Empire*, Transworld: London.

Palmer, R. and Parson, N. (eds.) (1983), *The Roots of Rural Poverty in Central and Southern Africa*, Heinemann: London.

Parker, G. (1988), *Europe in Crisis, 1598-1648*, Fontana: London.

Parsons, A. (ed.) (1948), *Seventy-five Years of Progress in the Mineral Industry, 1871-1946*, American Institute of Mining and Metallurgical Engineers: New York.

Payne, P. (1967), 'The Emergence of the Large-Scale Company in Great Britain, 1870-1914' in *Economic History Review*, Vol. XX, pp. 519-542.

Pollard, S. (1988), *Peaceful Conquest: The Industrialisation of Europe, 1760-1970*, Oxford University Press.

Powell, E. (1966), *The Evolution of the Money Market, 1395-1915: An Historical and Analytical Study of the Rise and Development of Finance as a Centralised, Co-ordinated Force*, Frank Cass: London.

Ramsey, P. (1972), *Tudor Economic Problems*, Victor Gollanz: London.

Reader's Digest Association Ltd. (1991), *Reader's Digest Atlas of the World*, London.

Reuff, J. and Hirsch, F. (1965), *The Role and the Rule of Gold: An Argument*, (Princeton Essays in International Finance, no. 47), Princeton Univ. Press: Princeton.

Richardson, P. (1977), 'The Recruuiting of Chinese Indentured Labour for the South African Gold Mines, 1903-1908' in *Journal of African History*, Vol. XVII, no. 1, pp. 85-108.

Richardson, P. and Van Helten, J. (1982), 'Labour in the South African Gold Mining Industry, 1886-1914' in Marks, S. and Rathbone, R. (eds.), *Industrialisation and Social Change in South Africa: African Class Formation, Culture and Consciousness, 1870-1930*, Longman: London, pp. 77-98.

_____(1984), 'The Development of the South African Gold Mining Industry, 1895-1918' in Economic History Review, Second Series, Vol. XXXVII, no. 3 (August 1984), pp. 319-40.

Rickard, T. A. (1895), 'Gold-Milling in the Black Hills, South Dakota, and at Grass Valley, California' in *Transactions of the American Institute of Mining Engineers*, Vol. XXV, Feb. 1895-Oct. 1895, pp. 906-929.

Rubin, I. (1979), *History of Economic Thought*; Ink Links, London.

Sampson, A. (1987), *Black and Gold: Tycoons, Revolutionaries and Apartheid*, Hodder and Stoughton: London.

Schmeisser, K. (1898), *The Gold Fields of Australasia*, Macmillan: London.

Sender, J. and Smith, S. (1986), *The Development of Capitalism in Africa*, Methuen: London.

Sennholz, H. (ed.) (1975), *Gold is Money*, Greenwood Press, London.

Service, R. (n.d.), *Collected Verses of Robert Service*. Ernest Benn Ltd: London.

_____(1986), *The Cremation of Sam McGee*, Kids Can Press Ltd: Toronto.

_____(1987), *The Shooting of Dan McGrew*, Kids Can Press Ltd: Toronto.

Skinner, E. and Robinson, H. (1915), *Mining Costs of the World*, The Maple Press: York, Penn.

Smith, G. (1943), 'The History of the Comstock Lode, 1850-1920' in *University of Nevada Bulletin* (Geology and Mining Series No. 37), Vol. XXXVII, no 3, 1 July, Nevada State Bureau of Mines.

Soetbeer, A. (1879), 'Edelmetall-Produktion und Wertverhältniss zwischen Gold und Silber seit der Entdeckung Amerikas bis zur Gegenwart', Ergänzungschaft Nr. 75 zu *'Petermanns Mitteilungen'*, Gotha.

Stern, F. (1977), *Gold and Iron: Bismarck, Bleichröder and the Building of the German Empire*, Allen & Unwin: London.

Temple, J. (1972), *Mining and International History*, Benn: London.

Tew, B. (1988), *The Evolution of the International Monetary System, 1945-1988*, Hutchison: London.

Ticktin, H. (1991), *The Politics of Race: Discrimination in South Africa*, Pluto Press: London.

Trebilcock, C. (1986), *The Industrialisation of the Continental Powers*, Longman: London.

Trewhela, P. (1970), *The Development of World Economy and the War in South Africa, 1890-1910*, unpublished MA Dissertation, University of Sussex.

_____(1986a), *The Problem of Subject in the Modern World: The Role of the Proletariat of Southern Africa: A Critical Analysis*, part of unpublished manuscript of PhD thesis.

_____(1986b), *The African Subject of Twentieth Century World Money-Dealing Capital*, part of unpublished manuscript of PhD thesis.

Van Helten, J. (1982), 'Empire and High Finance: South Africa and the International Gold Standard, 1890-1914' in *Journal of African History*, 23 (1982), pp. 529-48.

Van Onselen, C. (1976), *Chibaro: African Mine Labour in Southern Rhodesia, 1900-1933*, Pluto Press: London.

_____ (1982a), *Studies in the Social and Economic History of the Witwatersrand*, Vol. I: New Babylon, Ravan: Johannesburg.

_____ (1982b), *Studies in the Social and Economic History of the Witwatersrand*, Vol. II: New Neniveh, Ravan: Johannesburg.

Vicker, R. (1976), *The Realms of Gold*, Robert Hale: London.

Vickers, D. (1968), *Studies in the Theories of Money, 1690-1776*, Augustus M. Kelley: New York.

Vilar, P. (1976), *A History of Gold and Money 1450-1920*, New Left Books: London.

Walmesley, O. (1894), *Guide to the Mining Laws of the World*, Eyre & Spottiswoode: London.

Walmsley, J. (1979), *Macmillan Dictionary of International Finance*, Macmillan: London.

Warrington, J. (1968), *The Fortune Down Under*, Investment Publications of Australia: Alexandria, NSW, Australia.

Watermeyer, G. and Hoffenberg, S. (1932), *Witwatersrand Mining Practice*, Transvaal Chamber of Mines: Johannesburg.

Webster, E. (1978), *Essays in Southern African Labour History*, Ravan Press: Johannesburg.

Weinstein, A. (1970), *Prelude to Populism: Origins of the Silver Issue, 1867-1878*, Yale Univ. Press: London.

Wheatcroft, G. (1986), *The Randlords: The Men who Made South Africa*, Weidenfeld and Nicholson: London.

Williams, E. (1944), *Capitalism and Slavery*, University of North Carolina Press: Chapel Hill.

Williams, M. (1975), 'An Analysis of South African Capitalism: Neo-Ricardianism or Marxism?' in *Journal of the Conference of Socialist Economists*, London, Feb., pp. 1-28.

Willis, P. (1989), 'Trackless Mining in the Context of an Integrated Mining System' in *South African Mining: Coal, Gold and Base Metals*, May 1989, pp. 29-64.

Wilson, F. (1972), *Labour in the South African Gold Mines, 1911-1969*, Cambridge University Press.

Wise, E. (ed.) (1964), *Gold: Recovery, Properties and Applications*, D. van Nostrand: London.

Wolpe, H. (1972), 'Capitalism and Cheap Labour-power in South Africa: From Segregation to Apartheid' in *Economy and Society*, Vol I, pp. 425-456.

Collected works

LCW	V.I. Lenin Collected Works, Lawrence and Wishart: London.
MECW	Karl Marx, Friedrich Engels Collected Works, Lawrence and Wishart: London.
MEGA	Karl Marx, Friedrich Engels Gesamtausgabe, Dietz Verlag: Berlin, (1975+ edn.)
MEW	Karl Marx, Friedrich Engels Werke, Dietz Verlag: Berlin.

Periodical, newspapers, etc.

Cowell, A.: *The Decade of Destruction: Mountain of Gold*, Channel Four Television, London, 1 Oct. 1990.

Chamber of Mines Newsletter, Johannesburg.

The Economist, 1 November, 1975, London.

The Mineral Industry: Its Statistics, Technology and Trade, Annually from 1886 to 1908, The Scientific Publ. & Co: New York, London.

The Mining Manual, Annually from 1889 to 1899.

The Observer, 2. vii. 89.

Official and semi-official publications

ICE (1897), *Report of the Industrial Commission of Inquiry into the Gold Mining Industry*, Chamber of Mines, Johannesburg.

London Stock Exchange Yearbooks, 1894-1897.

Report of the Gold Production Committee (1918),*Gold Production of the British Empire*, HMSO, London.

The Statesman's Year Book, 1890-1905, London.

Transvaal Chamber of Mines Annual Report, 1902 plus Annexure.

United States Government, House of Representatives (1870), *Gold Panic Investigation Report*, Report no. 31, 1 March.

_____Senate (1879[1978]), *International Monetary Conference, 1878, Proceedings and Exhibits*, reprinted Arno Press, New York.

_____(1982), *Report to the Congress of the Commission on the Role of Gold in the Domestic and International Monetary Systems*, US Gov. Printing Office, Washington DC, March 1982.

Witwatersrand Chamber of Mines Annual Reports, 1889-1899.

ZAR Govt. (1895), *Hoofd van het Mijnwezen Rapport*, Pretoria.

Archival material

Marx, K.: *Geldwesen, Kreditwesen, Krisen*, Original Manuscript B-72, Karl Marx, Friedrich Engels Nachlaß, Institute of Social History, Amsterdam.

Rothschild Archives London, XI/62/23A, Letters of August Belmont and Co., New York to N.M. Rothschild and Sons, London.

Index

A
Abbysinia 43
abolition of slavery 4
abstract labour 29, 31, 129, 130, 176, 188, 198, 221, 233, 236-7
abstract wealth 33, 50, 54, 196
accumulation 32-39, 50, 84, 107, 131, 205
Act of Union 216
Adolf Goerz and Co. 161, 162, 163
Africa 4, 13
Africans 3, 150, 174, 202, 204-9, 212-17, 221-3, 225-7, 231, 233, 234, 237, 239, 249, 250
Alaska 4, 106, 203
Albu, G. 161, 162, 163, 206, 209, 210, 224, 226-7
Ally, R. 3
amalgam thefts 212
ancient Egypt 10, 11, 14, 109
Anglo-Boer War (*see* South African War)
Anikin, A. 2
antiquity 4, 10, 51, 84
apartheid 147, 240
Argentina 51
Ashanti 11
Association of Mines 161, 163
attributes of capital 85, 91, 93, 115, 123
attributes of gold (natural, social) 28-36
attributes of labour 6, 27, 28, 85, 115-6, 129-30, 194, 234
attributes of money 32-6, 94
Australia 4, 7, 11, 103-6, 110, 136, 138, 203, 239

B
backwardness 70, 121, 165, 202, 203
barter 36, 49, 64, 71, 82, 226
Berlin 159
bimetallism 136, 137
black miners 138, 161, 174, 176, 188, 211, 213, 223, 228-39, 249
Boksburg Tramway 227, 231
Bonacich, E. 5, 198, 215, 222, 223
Brakhan, A. 211
British Columbia 4, 56, 57, 121, 138
Buffelsdoorn Gold Mine 207
Bundy, C. 205-6, 227

C
cadastral boundaries 170
California 4, 6, 7, 11, 56, 57, 103-8, 110, 117, 138, 203, 214
Cape Colony 227
carat 124
Cariboo District 56
caste system 116

category (of abstraction) 4, 5, 7, 10, 77, 86, 139, 147, 189, 199, 214, 217, 218, 239
cattle 36, 95, 196, 208, 226
Cendrars, B. 16
centralised buying 168, 226
Chamber of Mines 92, 150, 160-3, 166, 175, 176, 180, 223, 237, 250
China 230
Chinese 6, 56, 57, 202, 214, 229-35
chlorination process 112, 127, 134
circuit of capital 79, 81, 83, 88, 89, 93, 154, 216, 217, 223, 226, 247, 252, 253
circulation 25, 28, 29, 33, 36, 39-44, 49, 50, 52, 54, 58-73, 78-80, 84, 91, 93, 94, 107, 124, 136, 155, 158, 191, 197, 200, 212, 225, 247
City of London 158, 162, 177
Clapham, J. 115
clearance system 67
closed compounds 188, 207, 209-12, 229-31, 235-6, 238, 250
coerced labour 15, 56, 131, 230, 233, 239, 250
coins 1, 42, 43, 54, 59, 60, 62, 78, 80, 93, 116, 118, 226
collusion 93, 149, 152, 153, 157, 250
Colorado 4, 117
commensurability 130
Committee of Agents 231
commodity 5, 6, 11, 25, 27-44, 49, 53-72, 77, 78, 79, 80, 83, 84, 86, 91, 92, 95, 101, 131, 151, 152, 157, 164, 174, 181, 189, 191-9, 214, 216, 220-22, 224, 236, 247, 252, 253
commodity production 5
competition 32, 71, 72, 84, 86-92, 108, 130, 139, 149-66, 172, 173, 175, 176, 179-82, 197, 214, 223, 227, 232, 236, 251
concrete labour 11, 13, 31, 34, 77, 129, 130, 134, 189, 190-92, 196

conglomerates 9, 10
Conquest, R. 13, 14
Conquistador 39, 104
constancy 10, 11, 25, 27, 31, 39, 122, 174
consumption 32, 35, 36, 41, 43, 44, 49, 50, 54, 58, 77-79, 82, 83, 86, 94, 95, 107, 122, 131, 163, 166, 199, 206, 225, 249
contradiction 6, 34-35, 50, 59, 62, 64, 65, 68, 71, 92-93, 163-5, 175, 194, 212, 221, 227, 253
Contribution to the Critique of Political Economy 5
convertibility 63, 64, 65, 66, 71
co-operation 11, 84, 106, 108, 149, 181
copper 117, 123, 182
Corner House 173
Coronation Syndicate 177
corvée 226
cost-reduction 165, 173
credit 27, 78, 201, 230, 253
Crown Reef Mine 203
currency 1, 2, 60, 61, 62, 63, 64, 65, 66, 67, 68, 69, 71, 72, 73, 78, 115, 155, 176
cyanidation process (*see* MacArthur-Forrest process)

D

De Beers Consolidated Mines Ltd 152, 160, 231
De Brunhoff, S. 2
deep-level mining 10, 161
Del Mar, A. 56, 132, 138, 239, 249
Denny, G. 211, 226
departments of production 95
desertions 56, 213, 231, 234
devaluation 33
digger 6, 50, 51, 52, 56, 72, 103, 107, 109, 111, 150, 196, 253
dividends 89, 90, 91, 117, 166, 176, 179, 180, 182
dollar 2
drilling 111

Dunning, J. 115
dynamite 14, 119, 120, 122, 126, 166, 175

E
economic history 25
economists 2
economy 3, 39, 49, 51, 52, 53, 62, 64, 67, 68, 77, 81, 82, 85, 86, 92, 93, 101, 105, 127, 128, 130, 132, 151, 165, 168, 190-92, 198-209, 213, 215, 223, 253
electricity 14, 120, 121, 122
empiricist 4, 155
Engels 2, 134 137 248
Europe 39, 40, 104, 131, 132, 177, 203, 204, 205, 224, 228, 233
Evans 205
evolution of money 4
exchange value 7
explosives (*see* dynamite)
expression 52
extraction 126
extractive industries 12, 70

F
factors of production 92, 130, 154, 157, 160, 163, 165, 168, 173
feudal 83
finance 2, 154, 157, 158, 164-6, 169, 173, 176, 177, 182, 219
Finance/Mining House 167, 169, 170
financial markets 91, 157, 158, 162, 163, 167, 169, 170, 172, 173, 177, 181
fineness 105, 124, 125
Finley, J. 134
First World War 2
FitzPatrick, J. 212
forced labour (*see* coerced labour)
Forty-Niners 11
Fossickers 11
France 219, 220

Frankel, S. 167
Fraser, M. 177
free labour 7, 37, 131-2, 138, 200, 229, 230, 234-5

G
G. and L. Albu and Co. 161
Georgia 104, 108
geo-sciences 179
Gilliani 28
Goerz, A. 162, 163
gold as capital 5, 77-9, 95, 247-8, 253
gold as money 5, 27-41, 51, 55, 62, 71, 95, 181, 196, 220, 247, 248, 253
gold as use-value 5
gold as use-value I 49, 95, 247, 248, 253
gold as use-value II 50, 77, 195
Gold Coast 11
Gold Law 212
Goldmann, C. 226, 238
gold particles 9, 10, 11, 32, 112, 133
gold-producing capital 1, 81, 85, 86, 220, 247, 251, 253
gold-producing labour-power 1, 27, 55, 58, 67, 73, 77, 95, 147, 247, 251
gold rush 12, 15, 16, 103, 104
gold standard 2, 39, 67, 68, 154-5
gold-washing 14, 15
Gool, S. 3, 233, 249
grade 4, 16, 70, 112, 125, 126, 154, 167, 171, 178-80, 202, 204
ground sluicing 109, 110
Group System 149, 150, 157, 159, 165-73, 182, 183, 251
Grundrisse 2, 5, 163, 193, 247
guild 153

H
Hall, W. 203, 214, 215
'hard money' 25

Harris, P. 205, 209, 223
Hay, E. 210
Hay, J. 222
hedonist 32
Hirson, B. 138
hoard 27, 32, 33, 39, 42, 43, 67, 131, 182
Hobson, J. 2, 129, 212-13, 225-7
houseboy 228, 232, 233
human labour in the abstract 29, 189, 191, 192, 193, 195
Hungary 175
hydraulic method 57, 109

I
ICE 203, 205, 206, 207, 209, 210, 212, 214, 215, 222, 224, 226, 227, 228, 238
imperialism 2, 4, 204
indentured labour 6, 129, 202, 208, 214, 229-235
India 4, 131, 239
Indian 43, 56, 57, 239
industrial revolution 4, 7, 101, 113, 115, 118
inflation 2
Innes, D. 2, 3, 149-64, 170, 172, 214, 215, 248
interest rate 67
inventions 13, 14, 103, 118, 179
investors 88, 112, 158-9, 162, 169, 170-73, 177
Investor's Guardian 167, 168, 177, 180

J
J. B. Robinson & Co. 161
jackhammer 119
Jacob, W. 32, 38, 138
Jameson Raid 162
Japan 175, 230
Jastram, R. 27, 38
Jeeves, A. 177, 182, 205, 208, 231
Jennings, S. 210
Jennings, H. 228

jewellery 41, 42, 43, 59, 124
Jews 43
Johannesburg 160, 177, 203, 228, 231-3, 239, 249
Johnstone, F. 149-60, 164, 172, 182, 183, 234
joint-stock company 4, 6, 15, 72, 88, 110, 111, 116, 159, 166, 173, 176

K
'Kaffir' 177, 178, 207, 208, 209, 211, 227, 233
'Kaffir Boom' 227
Kallaway, P. 235
Katzen, L. 3, 174
Kemp, T. 115
Kimberley 150, 152, 153, 160, 188, 210, 211
Klondike 4, 103, 106
Kolyma 14
Kubicek, R. 162
Kuwait 220

L
Labour Reserves 188, 198, 213, 216-18, 220-1
Land Acts 188
Landes, D. 115
Lanning, G. 236
Latin America 15
laws of capital 6
laws of money 6
Legassick, M. 204
Lenin 2, 196
Lightning Amalgamator 112
Lightning Creek 121
locations 207, 209, 210-13, 227-8
lodes 6, 8, 9, 11, 15, 16, 84, 103, 104, 106, 110-16, 131-5, 137, 155, 188, 202, 229
London Stock Exchange 176
long-Tom 104, 110
low-value, high-productivity 7, 86, 203

low-value, low-productivity 7, 86
Luso-Gaza War 209
luxuries 94, 95, 136, 224

M

MacArthur-Forrest process 122, 127, 134, 170, 179
machinery 12, 104, 107, 111, 118, 121-2, 134-5, 179, 206
Maghreb 104
Mandel, E. 2
manufacturing 118, 188, 233
market 86, 87, 88, 104, 123, 130, 136, 138, 151-7, 160, 164, 165, 173, 174, 177-9, 181, 182, 188, 198, 201, 204, 206, 213, 215, 220-8, 232, 236
Marks, S. 177
Massachusetts 203
Masters and Servants Laws 235
Matabeleland 12
means of payment 27, 65
means of production 80-81, 84, 87-88, 122, 137, 163, 166, 188, 216-9
measure of value 1, 28-31, 52, 58-63, 72, 106, 110, 137, 155
medium of circulation 27, 28, 65, 69, 78
MEGA 28, 248
merchant's capital 64, 77, 79
mercury 15, 105, 122, 127, 133, 134, 135
metallurgy 10
method (of abstraction) 2, 5, 6
Mexico 111
Middle Ages 4, 104
migrant labour 129, 187, 188, 209, 211, 216-222
mine owner 16, 209, 212-3, 224-9
mines 2, 7-14, 51, 56, 57, 87, 95, 107-12, 117, 119-23, 131, 134-8, 150, 156, 159-82, 187-93, 198, 203-14, 221-3, 227, 228-39, 249, 250
miser 32

money-commodity 5, 28, 36-44, 49, 60, 68, 73, 77, 81, 86, 95, 101, 125, 131, 137, 138, 154, 181, 196, 221
money dealer's capital 77
money economy 27
monopoly 70, 81, 92, 130, 149-61, 166, 169, 175, 182, 251
monopoly capitalism 4, 149-52, 162, 204
monopsony 92, 157, 158, 160
Moroccans 219
multinational corporation 115
Mussollini 43

N

Natal 227, 229, 230
native gold 8, 133, 134
Nazis 43
necessary labour-time 16, 39, 40, 43, 50-8, 84, 106-7, 110, 131, 135, 155, 201, 216, 222, 225-6
neo-classical economics 2, 5
Nertschinsk 239
Netherlands South Africa Railway Company 166
New South Wales 107, 111
New World 39, 40, 101
New Zealand 4
nineteenth century 1, 2, 6, 8, 10, 13, 25, 30, 69, 70, 73, 77, 103, 104, 110, 115-38, 150, 151, 158, 170, 202, 209, 213, 247
North America 131, 135, 203

O

oligopoly 153, 156, 160, 161, 173
oligopsony 160
ore reduction 123
over-work 7, 49, 51, 84

P

Pakistanis 220
Paramillo 51

Pass Law 208
Payne, P. 115
Pearson, P. 235
peasants 43, 206, 227
personal bondage 104
Phillipinos 220
Phillips, L. 205
placers 6–11, 15, 16, 31, 37, 38, 81, 84, 103-12, 116, 131, 133, 137, 155, 188, 202
planned production (*see* rational production)
plunder 37-40, 81, 84, 101, 103-4, 131
pneumatic tools 14, 121
political economy 1, 2, 3, 7, 11, 16, 25, 44, 52, 134, 147, 173, 182, 198, 211, 230, 233, 235, 253
Pollard, S. 115
Pondo 95
Portuguese 209
Potosi 37
Pretoria 111
price 39, 40, 41, 54, 58-63, 66, 68, 70-72, 87, 88, 90, 91, 156-60, 165, 173, 180, 182, 188, 201, 214, 215, 219, 222, 224-6, 236
prisoners 14
private capital 72, 88, 90, 93
product 5, 6, 31, 34, 35, 41, 42, 44, 50, 51, 53, 55, 58, 70, 71, 72, 79, 81, 82, 84, 86, 87, 92, 96, 110, 115, 120, 192, 195, 196, 225
productivity of labour 30, 37, 39, 53, 57, 62, 69, 70, 83, 86, 87, 92, 94, 95, 101, 115-27, 163, 173, 181, 182, 201-3
profitability 85, 115, 127
proletarianisation 129, 176, 188, 205, 213, 229

Q

quantity vis-à-vis quality of money 32, 34-35, 38-40
quartz 9, 16, 110, 111, 135

R

railway rates 166
Ramses 12
Randlords 182
rate of exchange 38, 62-4, 66-9, 71, 202
rational production 87, 153, 157, 159, 163-5, 168-76
realisation 33, 66, 69, 116, 129, 156
reduction 126, 127
refining 123, 127, 133
relations of production 109, 110, 251
relative form (of value) 59, 94
rent 70, 153, 225, 226, 253
Richardson 3, 104, 177, 179, 205, 212-8, 230, 233, 235
Robinson, J. 161, 162, 163, 212
Rome 15
Rothschild Group 152
Russia 13, 15, 56, 131, 239

S

Sahara 104
securities markets 176
Seti 14
settler economy 219
shaft 11, 15, 51, 104, 109, 117, 118, 119, 120, 122, 170, 180
Siberia 4, 14, 56, 110
Siculus 49, 51, 84, 137
silver 2, 27, 28, 33, 36, 37, 40, 42, 44, 49, 51, 84, 104, 111, 123, 135, 137, 193
simple labour 189, 190, 196
slavery 7, 37, 51, 128, 131, 132, 202, 229, 230, 235, 239, 250
social capital 71, 90-3, 115-6, 119, 123, 129, 136, 153, 158, 163, 173
social economy 28
social form 6, 13, 25, 27, 40, 41, 53, 56, 72, 85, 130, 190, 195-7, 199, 202, 253, 254
social labour 34, 64, 130, 190, 192, 199

social relations of production 108
socialism 196, 220
société anonyme, 115
South African War 199, 206, 212, 227, 234, 236
South America 4
South Sea Bubble 115
southern Africa 5, 175, 187, 189, 205, 217, 236, 239
speculation 63, 91, 115, 117, 176
Stalin 13, 14
standard of living 50, 57, 86, 200-3, 220, 223, 228
standard of price 58-61, 63, 67, 69, 71, 72, 101
state 4, 15, 56, 62, 63, 66, 68, 72, 83, 91, 128, 129, 157, 166, 176, 205, 208, 213, 224-6, 240, 250
steam 13, 118, 120, 121, 122, 123, 228
stock exchange 115-7
stores 165, 166, 173, 237
subsistence 43, 81, 198, 199, 200, 206, 212, 216, 221, 225
substance of value 52
supply and demand curves 25
surface work 7, 8, 10, 104, 133
surplus product 81, 206
surplus-value 1, 7, 50, 51, 70, 71, 79, 83-91, 94, 130, 149, 150, 153, 157, 163, 166, 173-4, 182, 216, 228, 252
Sutter, J. A. 16

T
Tacquah 13
Tati 12, 111
tax 209, 225, 226, 227
technology 11, 12, 14, 50, 57, 70, 84, 104-27, 171-2, 182
Theories of Surplus Value 93
theory of gold 1, 3, 25, 81, 93, 247, 253
theory of luxuries 93
theory of money 93

Ticktin, H. 4, 33, 87, 129, 130, 181, 187, 189, 191, 198, 218, 233, 236
token 60-72, 78, 85
tramming 120, 121, 167
Transvaal 227, 230
Transvaal-Moçambique Agreement 209
Trapido, S. 177
Trebilcock, C. 115
Trewhela, P. 3, 87, 115, 116, 181, 187, 189, 194-8, 220, 248

U
underground 7-8, 11, 14, 16, 38, 57, 111, 118, 120-22, 134, 167, 170, 212, 231
unfree labour (*see* coerced labour)
unit of labour 30, 31
United Kingdom 165
United States 2, 56, 117, 128, 136, 203
Unity Movement 187
universal equivalent 41, 60, 61, 193, 194, 197
universal rise in prices 40
universal use-value 41
use-value I 41-4, 49, 54, 77-8, 194-7
use-value II 41-4, 49, 54, 77-8, 137, 194-7
USSR 10
usurer's capital 77

V
value of gold 1, 14, 29, 31, 38-40, 52-4, 58-60, 69-72, 78, 85, 92, 109, 110, 125, 131, 133, 137, 155
value of labour-power 1, 4, 5, 49, 50-3, 55, 57-8, 62, 68-9, 84, 86, 94-5, 111, 133, 138, 147, 174, 198-202, 214-9, 222, 227, 235, 237, 239, 249, 251-3
value revolution 99
value-for-itself 37

value-ratio 1, 38, 135-7
Van Helten, J.-J. 3, 104, 179, 205, 212, 218
Van Onselen, C. 212, 213, 228, 231, 236
variable capital 1
Venezuela 228
Victoria 111
Victorian Britain 12
Virginia 4, 110, 112, 117

W

wage-labour 5, 6, 7, 50, 53, 54, 83, 85, 110, 116, 122, 128, 129, 132, 189, 197-202, 205, 213, 221, 226, 228, 230, 235, 236, 249, 250, 252
wages 4, 57, 94-96, 108, 138, 161, 166, 174, 175, 182, 188, 199, 200, 201, 204, 206, 212-8, 221-8, 230, 233, 237, 249-52
wealth 32, 35, 38, 39, 41, 42, 44, 49, 50, 52, 54, 55, 83, 95, 115, 151, 194, 201, 209, 224
Webster, E. 204
Werner Beit 162
Wertpapier 115
white miners 4, 56, 165, 174, 175, 187, 199, 202, 211, 214, 216, 218, 222, 223, 228, 249, 250, 251, 252

Williams, M. 3, 87, 93, 181, 189-99, 212, 215, 218, 221, 248, 250, 251
Wilson, F. 164, 172, 174, 177
Witwatersrand 6, 10, 103, 119, 127, 134, 149, 150, 152, 159, 166-73, 176, 178, 179, 182, 188, 202-3, 207-10, 214-6, 227-31, 235, 238-9, 249
Witwatersrand Native Labour Association 175
Wolpe, H. 3, 198, 216, 218, 221
working class 3, 4, 187, 200, 218
world 39, 67, 77, 153, 173
world economy 2
world gold production 105

Y

Yenesei 56
yield (gold) 3, 11, 14, 16, 107, 111, 117, 126-7, 135, 167, 169, 178-9
yield (surplus-value) 130, 216

Z

Zimbabweans 220
Zuid Afrikaansche Republiek (*see* Transvaal)

DATE DUE

UPI 261-2505 G   PRINTED IN U.S.A.